The Domestic Cat: Bird Killer, Mouser and Destroyer of Wildlife
Means of Utilizing and Controlling the Domestic Cat

by Edward Howe Forbush

with an introduction by Jackson Chambers

This work contains material that was originally published in 1916.

This publication is within the Public Domain.

*This edition is reprinted for educational purposes
and in accordance with all applicable Federal Laws.*

Introduction Copyright 2018 by Jackson Chambers

COVER CREDITS

Front Cover
Cat and Mouse by User: Lxowle (User made.)
[CC BY-SA 3.0 - https://creativecommons.org/licenses/by-sa/3.0]
or
[GFDL - http://www.gnu.org/copyleft/fdl.html],
via Wikimedia Commons

Back Cover
Yawning Norwegian Forest Cat by User: Mattes (Own work)
[Public domain],
via Wikimedia Commons

Research / Resources
Wikimedia Commons
www.Commons.Wikimedia.org

Many thanks to all the incredible photographers, artists,
researchers, and archivists who share their great work.

PLEASE NOTE :
As with all reprinted books of this age that are intended to perfectly reproduce the original edition, considerable pains and effort had to be undertaken to correct fading and sometimes outright damage to existing proofs of this title. At times, this task can be quite monumental, requiring an almost total rebuilding of some pages from digital proofs of multiple copies. Despite this, imperfections still sometimes exist in the final proof and may detract slightly from the visual appearance of the text.

DISCLAIMER :
Due to the age of this book, some methods or practices may have been deemed unsafe or unacceptable in the interim years. In utilizing the information herein, you do so at your own risk. We republish antiquarian books without judgment or revisionism, solely for their historical and cultural importance, and for educational purposes.

Self Reliance Books

Get more historic titles on animal and stock breeding, gardening and old fashioned skills by visiting us at:

http://selfreliancebooks.blogspot.com/

introduction

Here at **Self-Reliance Books** we are dedicated to bringing you the best in *dusty-old-book-knowledge*. We are so happy to bring you another old book on Cats.

This book is *NOT* for the Cat-Lover! I mean this when I say it. In fact, it's for people who are not awfully fond of cats at all!

This special edition of **The Domestic Cat : Bird Killer, Mouser and Destroyer of Wildlife – Means of Utilizing and Controlling the Domestic Cat** was written by Edward Howe Forbush, and first published in 1916, making it over a century old.

The book, which has a focus on controlling and utilizing the cat as a tool, has sections on *Origin, History, Fitness, Character and Intelligence, Natural Enemies of the Cat, Food, The Economic Value of the Cat, Means of Controlling the Cat,* and more.

Again. this book is *NOT* for Cat-Lovers, but is a great read for all those who may want to utilize the cat as a tool to control pests, perhaps around a farm or ranch, or other country or remote property.

~ *Roger Chambers*

State of Jefferson, March 2018

Source: Wikipedia.com
British Longhair Black Silver Shaded
July 2008
Creative Commons Attribution
3.0 Unported, 2.5 Generic, 2.0 Generic and 1.0 Generic license.

Source: Wikipedia.com
CC BY-SA 3.0
File:Russian Blue 001.gif
Created: 12.05.06

PREFATORY NOTES.

Questions regarding the value or inutility of the domestic cat, and problems connected with limiting its more or less unwelcome outdoor activities, are causing much dissension. The discussion has reached an acute stage. Medical men, game protectors and bird lovers call on legislators to enact restrictive laws. Then ardent cat lovers rouse themselves for combat. In the excitement of partisanship many loose and ill-considered statements are made. Some recently published assertions for and against the cat, freely bandied about, have absolutely no foundation in fact. The author of this bulletin has been misquoted so much by partisans on both sides of the controversy that in writing a series of papers on the natural enemies of birds it has seemed best, in justice to the cat and its friends and foes, as well as to himself, to gather and publish obtainable facts regarding the economic position of the creature and the means for its control.

The first publication of the State Board of Agriculture that referred particularly to the natural enemies of birds was a special report on the "Decrease of Certain Birds and its Causes," published in the fifty-second annual report of the Board in 1904. A paper on the English sparrow appeared in the fifty-eighth annual report, and one on the starling in the fifty-ninth. These two papers, revised and enlarged, have been republished in 1915 as circulars 48 and 45 respectively. Bulletin No. 1 of the present series, already in its second edition, treats of the rat as an enemy of mankind and birds, and deals with the means of suppressing it. The rat, although of less importance than the cat as a bird killer, was considered first, for people who intend to dispose of their cats need first to know how to rid their premises of rats.

This paper has been written in the hope that it will interest and inform not only cat lovers and bird lovers, but that large part of the public whose attention is engaged at times by both cats and birds. An attempt has been made to avoid unnecessary scientific verbiage and to set forth the facts plainly and convincingly.

The Houghton-Mifflin Company of Boston and the Lothrop, Lee & Shepard Company of New York have given permission respectively to quote from Miss Repplier's charming volume "The Fireside Sphinx" and from Miss Winslow's "Concerning Cats." Charles Scribner's Sons have granted a similar privilege regarding Shaler's "Domesticated Animals."

Mrs. Huntington Smith, president of the Animal Rescue League of Boston, has kindly proffered the use of much material that she has gathered from friends of the cat.

Edward N. Goding, Esq., has read that portion of the manuscript devoted to the cat in law, and has given valuable suggestions.

Mr. Alfred Ela has contributed the use of all his notes and clippings relating to the subject.

The line drawings are from the pen of Mr. Walt F. McMahon.

The author has received very material aid from the National Association of Audubon Societies and is indebted to many authors and to a host of correspondents, much of whose material could not possibly be utilized within the limits of this bulletin; nevertheless, it has been given due weight in arriving at conclusions.

CONTENTS.

	PAGE
Introduction,	7
Origin,	8
History:	10
The cat in Egypt,	10
The cat in Asia,	13
The cat in Europe,	13
Fitness, Character and Intelligence:	14
Cruelty of the cat,	15
The cat compared with the dog,	16
Independence of the cat,	17
Affections of the cat,	17
Fecundity of the cat,	19
Natural enemies of the cat,	19
Numbers of cats:	19
Great numbers of vagrant cats in cities,	20
Numbers of vagabond or wild house cats in the country,	22
Cats abandon owners,	25
Owners abandon cats,	25
Cats unfed by owners,	26
Habits,	27
Food:	28
Vegetal food of the cat,	28
Animal food of the cat,	28
Destruction of insectivorous birds by cats,	29
The cat a birdcatcher in ancient times,	30
The cat a birdcatcher in modern times,	31
Birds cut by claws of cats may die,	33
Cat poaching for owner,	33
Active and intelligent birdcatchers,	34
Cats enticing birds,	34
Numbers of birds killed by cats,	34
Cats *versus* spraying trees,	35
Bird slaughter by cats,	35
Young birds the chief sufferers,	37
Statements from people in the country,	40
Cats allowed to roam at night,	41
Correspondents report many birds killed,	42
Number of birds killed per day, week, month and year,	43
Number of birds killed in various States,	44
Destruction of game birds by cats,	45
Bobwhites,	45
Ruffed grouse,	46
Heath hens,	47
Pheasants and partridges,	47
Snipe, woodcock and other game birds,	48
The cat on the game preserve,	48
Number of observers reporting game birds killed,	49
Destruction of poultry and pigeons by cats,	49
Chickens,	49
Young turkeys,	51
Bantam fowls,	51
Full-sized fowls,	
Pigeons or doves,	
Cats eating eggs,	
Extermination of island birds by cats,	

	PAGE
Food — *Concluded.*	
Animal food of the cat — *Concluded.*	
Expert opinions on the cat's destructiveness to birds,	58
Destruction of mammals and lower animals by cats,	61
Squirrels,	61
Hares and rabbits,	61
Moles and shrews,	62
Rats and mice,	63
Bats,	68
Reptiles and amphibians,	68
Fish,	68
Crustaceans and mollusks,	69
Insects,	69
The economic value of the cat:	70
Economic value of weasels,	70
Economic value of squirrels,	70
Economic value of hares or rabbits,	71
Economic value of moles,	71
Economic value of shrews and bats,	71
Economic value of amphibians and reptiles,	72
Economic value of birds,	73
Species of wild birds reported killed by cats,	74
Cats and insects increase,	76
Injury by insect pests,	76
Insect pests eaten by birds,	77
Number of insects eaten by birds,	78
Birds save trees and crops from destruction,	79
Inutility of the cat,	80
Animal substitutes for the cat,	81
Is the cat a disseminator of disease,	82
Parasitic diseases,	84
Infections from cats' claws and teeth,	84
Tetanus or lockjaw,	84
Rabies or hydrophobia,	85
Septicæmia or blood poisoning,	86
Means of controlling the cat:	87
Catproof fence,	88
Killing the guilty cat,	90
Confining or tethering the cat,	90
Keeping the cat indoors at night,	92
Feeding the cat,	92
Belling the cat,	92
Cat guards,	93
Keep only white cats,	94
Air guns, torpedoes, etc.,	94
Electrocution,	94
Dogs,	94
Training the cat not to catch birds,	95
To prevent cats killing chickens,	97
Legislation for the control of the cat,	97
Methods of taking and killing stray or feral cats,	100
Legal rights of the cat,	102
Recapitulation and conclusion,	106
List of those who contributed information,	109

THE DOMESTIC CAT.

INTRODUCTION.

The cat, of all animals, is in some respects the most intimate companion of man. It is more closely identified with indoor life and the cheerful domestic hearth than is any other animal. It is, as St. George Mivart says, "the inmate of a multitude of humble homes in which the dog has no place."

Its independent character and its graceful, quiet movements appeal particularly to women. Its elegance of form, beauty of coloring, daintiness of habit, and, above all, the delightful, playful activity of its young make it a welcome fireside companion throughout the civilized world, and the playmate of innocent children in countless happy homes. It is considered useful inasmuch as it tends to keep down the undue increase of rodent pests. Nevertheless, it leads a dual existence. "The fireside sphinx," the pet of the children, the admired habitué of the drawing-room or the salon by day, may become at night a wild animal, pursuing, striking down and torturing its prey, frequently making night hideous with its cries, sneaking into dark, filthy, noisome retreats, or taking to the woods and fields, where it perpetrates untold mischief. Now it ravages the dovecote; now it steals on the mother bird asleep on her nest, striking bird, nest and young to the ground. In the darkness of night it turns poacher. No animal that it can reach and master is safe from its ravenous clutches.

In justice to the cat it should be said that it cannot be blamed for following the natural propensities of the *Felidæ*, the carnivorous family of mammals to which it belongs. Man brought it to this country, and the disturbance of the balance of nature caused by its introduction is man's fault, and occurs because he failed to control his own pet and protégé. We are more to blame than the cat for its wide-roaming, bird-and-game-killing propensities. Many cats naturally are indolent and sedentary, and would not stray far from their homes unless driven by necessity, but the neglected one must bestir itself to live. Abandoned or deserted by human friends, often expected to hunt most of its and allowed to shift for itself, it must appease by its own efforts the hunger due to wandering, fighting and exposure. Many people express the belief that it is "a poor cat that cannot pick up its own living." Some never feed their cats, and we need not wonder that puss, neglected and spurned, becomes by necessity a scourge to wild life.

The cat is the only domestic animal which is not usually regarded as property under the law, and which is neither fully restrained nor protected by it, also the only one that commonly is allowed by its owner to run wild and get its own living. This, however, is the lesser evil. The greater lies in the fact that hundreds of thousands of cats, deserting their owners or deserted by them, have reverted to the wild state, bred in the woods, and the numbers of their progeny have increased until they have become such a menace to small game, insectivorous birds and poultry that some method of repressing them must be found. The situation has become so serious that the legislators of many States have been asked to consider measures for the repression of these nocturnal marauders.

In recent years, some evidence has been adduced in support of the claim that the cat disseminates disease, particularly among children.

The object of this bulletin is to discuss the origin, history, character, habits and economic position of the cat, and to consider how its beneficial habits may be fully utilized and its injurious habits minimized.

ORIGIN.

Mivart says that it seems probable that the Mammalia, which of course includes the cat, descended from some highly developed "somewhat reptile-like batrachian of which no trace has been found."

The origin of the domestic cat is not definitely known, but the beginning of its association with man and his home falls within historic times. All histories of ancient nations go back to a time when they had no cats. No trace of the house cat has been found among the early nomadic tribes. The Swiss lake dwellers of the Stone Age had no pet cats, although they hunted and ate a wild species. The Indo-Aryans of the Vedic Age had none. Ancient Greece and Rome were without them. The earlier records of civilization make no mention of the cat, nor is it represented as a domesticated animal on any of the most ancient monuments or works of art that have been discovered. The Bible omits it, but it is spoken of once in the Apocrypha. Some

Hebrew scholars, however, believe that the animal there referred to is the jackal. Even in Egypt, where the cat appears to have been first tamed and where it became an object of worship, its domestication seems to have been comparatively late. Everything points to the probability that the cat was domesticated originally in Africa. African cats are easily tamed, while those of other countries are said to be more savage and do not so readily lend themselves to domestication.

The cat appears to have come to the front as a domestic animal about the period of the twelfth dynasty in the "Land of Cush," after the conquest of that country. It seems probable,

Egyptian hunting cat, *Felis maniculata*. An ancestor of the domestic cat.

then, that this little Cushite was derived from the wild Kaffir cat, *Felis caffra*, or from *Felis maniculata*, which is a native of Nubia and the Sudan. Cat mummies from Egypt have been considered to belong to this species, but naturalists differ regarding the identification, and Blainville distinguishes three species among cat mummies, *Felis caligata*, the Egyptian cat (which is identical with *F. maniculata*), *F. bubastis* and *F. chaus*, an Asiatic species. Two of these species are found still, both wild and domesticated, in Africa. Ehrenberg, however, considers all the cat mummies that he examined as remains of the Abyssinian wild cat, *F. caligata*. Temminck, Pallas and Blyth conclude that the domestic cat, *Felis domestica*, is a result of the interbreeding of many species, and as there are many small wild cats in various parts of the world, and as *Felis domestica* breeds freely with *Felis catus*, the common wild cat of Europe, there seems to be a probability that the domestic cat is the product of many species.

Since writing the above I have devoted some attention to the probable origin of *Felis domestica*, and am now inclined to agree with Dr. D. G. Elliot in the belief expressed in his monograph of the Felidæ that *F. maniculata* and *F. caligata* are practically identical with *F. caffra*. It is well to keep in mind the fact that many closely allied forms which have been described as species or races may have no real basis in nature, except as they have emanated from the gropings of the human intellect. Probably all the members of this group of closely related African cats described under different names are identical with or were derived from *F. caffra*. According to Elliot, this widely distributed form seems to vary in color from dull yellowish to dark gray. It shows markings somewhat similar to the common tabby, but less numerous, and has a blackish phase also. Its variations in color include practically all those of the domestic cat, except such as are the product of domestication. Its appearance is much like that of the domestic cat, except that it seems somewhat slimmer than the usual form of the household pet. Anatomically it is much the same, if we allow for the changes produced by domestication. The sparse markings of this species may not account for the numerous ones of the domestic tabby, but these may have been produced centuries ago in Europe by many crossings with the well-marked wildcat *F. catus* when wildcats were numerous there and the domestic cat had not become common.

The cat certainly was domesticated in Egypt at least thirteen hundred years before Christ. One of the earliest representations of the cat with man is a statue of King Hana, probably of the eleventh dynasty, with his cat Bouhaki between his feet. References to the animal, found on monuments, appear in written rituals of the eighteenth dynasty, about 1500 B.C. Hieroglyphic inscriptions which go back to 1684 B.C., and some probably as far back as 2400 B.C., mention the cat. The earliest known pictorial representation of puss as a domestic pet is shown on a tablet of the eighteenth or nineteenth dynasty (about 1500 to 1638 B.C.) now at Leyden, where she is represented seated under a chair.

HISTORY.

The Cat in Egypt.

A full history of the cat in domestication would make an absorbing tale. In Egypt she sat in the seats of the mighty. She was dedicated to woman and to Isis or the moon, and possibly to the sun also. Plutarch says that the image of a female cat was placed at the top of the sistrum as an emblem of the lunar orb. Horapollo asserts that the cat was worshipped in the temple of Heliopolis, sacred to the sun. Some scholars claim to have found evidence that one sex was believed to be emblematic of the moon, and that the other was symbolic of the sun. Such homage was paid the animal possibly because its eyes change the form and size of their pupils with the waxing and waning of the orbs of day and night, and become more brilliant when the moon is full.

A cat-headed goddess appears in the temples of Egypt, known as Bast, Pasht, Sekhet, Pasche, Tefnut or Menhi, believed by some to have been the Diana or hunting goddess of the Egyptians. She is referred to by others as the goddess of love or pleasure. The cat well might be chosen to represent both Diana and Venus. This goddess, known to the Greeks as Bubastis, seems to have antedated the deification of the cat, and to have been a lioness goddess until the cat was domesticated, when the deification of the king of beasts apparently was forgotten, and the "little lion" of the fireside took its place as an object of veneration.

Egyptian cat goddess.

From the twelfth dynasty onward pussy seems to have become a precious jewel — a fetish of the Egyptian people. The valley of the Nile was then a great grain-growing region, and Egypt the granary of the ancient world. No doubt the utility of the cat in catching rats and mice appealed to the Egyptians, but this was merely incidental, and no adequate reason for the exceeding veneration with which cats were treated.

The extreme reverence, affection and solicitude displayed by the people of Egypt for this animal are illustrated by historic tales of the ancients which seem incredible in the light of the twentieth century. The law forbade the sinful killing of a cat. The city of Bubastis, now in ruins, between the arms of the Nile and above the present town of Benha-el-Asl, was dedicated to cats and cat worship. Bubastis was built in the time of Thothmes IV, about 1500 B.C. Herodotus records the pilgrimage of seven hundred thousand people to this city in one year, and asserts that the lives of cats were held so sacred that when a fire took place, and an impulse to rush into it seemed to possess the felines, the Egyptians occupied themselves with keeping them away from the burning building and neglected to quench the fire. In spite of all this tender solicitude some cats escaped and cast themselves into the flames, amid the wild lamentations of the bereaved and horrified Egyptians. All members of any family bereaved by the death of a cat had their eyebrows shaved off, and the sacred animal was embalmed and then buried at Bubastis.

Bronze statuette of the cat of Bubastis.

No Egyptian dared run the risk of injuring a cat. There is a tradition repeated by the old historians regarding Cambyses, the Persian king, who attempted to take the town of Pelusium but was beaten back by the Egyptians. The tale runs that he then gave living cats to the soldiers in the front ranks of his army, which they used as shields, and the Egyptians retired and gave up the town without striking a blow. Diodorus says that a Roman who killed a cat by accident in Thebes was almost torn to pieces by the infuriated populace.

The exportation of cats was prohibited. An Egyptian commission searched the Mediterranean countries to buy and bring back, if possible, every cat which had been taken out of Egypt.

The temples of Bubastis, Beni Hassan and Heliopolis were sacred retreats of the deified animal, but that of Bubastis was the "fairest in all Egypt." There the sacred cats were robed, pampered and worshipped during life. There their necks and ears were hung with jewels and ornaments of gold. There they "drowsed and played in the shadows of mighty temples," and there their remains were tenderly and reverently preserved after death. Mummies of cats that had lived in the temple of the Goddess Pasht at Bubastis were greatly venerated by the people, and their tombs contained great numbers of gold ornaments bearing the same letters as those found in the mausoleums of Egyptian kings. Cat mummies were wrapped in fine linen like that in which the remains of kings were swathed.

"How now are the mighty fallen!" In recent years, great cat burial places have been rifled of their sacred deposits and the bones used to fertilize Egyptian fields, or prepared and shipped abroad, to be sold at $15 a ton as fertilizer.

Outside of Egypt, with its pictorial art, mummies and inscriptions, the records of the early history of the cat are few. Little is known about its place in the homes of men between the time of the latest Egyptian records and about 260 B.C. when it appears as already established in Greece and Rome.

The Cat in Asia.

About 400 B.C. the cat is referred to in Chinese records as a wild animal, and does not appear to have been tamed in China until after the beginning of the Christian era. It appeared also in Persia and India, but the exact date of its first appearance in domestication there is one of the mysteries of the past, and whether it came there from Egypt and interbred with native types or was domesticated from native species alone cannot be determined. All long-haired cats, however, are believed to have come from the East, and seem to have had a common origin in Pallas' cat (*Felis manul*).

The Cat in Europe.

Some authorities assert that the cat came to Europe from Cyprus, others that it was introduced from Egypt. Diodorus says that hunters carried it away captive from Numidia to decadent Greece. Whatever may be the facts, its former glory had departed. In Greece and Rome it was little honored and less worshipped, but was tolerated and valued because of its ability as a mouser. Apparently it was disseminated slowly through Europe. There seems to be no proof of its domestication in Great Britain or France before the ninth century. Although its utility had been recognized early it soon became a beast of ill repute, — a reputation which followed it for centuries. Its cold temperament, nocturnal habits, flaming eyes and horrible night cries resulted in its becoming the victim of superstition. It was classed with devils, witches, sorcerers, owls, bats and the spirits of sin and darkness, and in the dark and middle ages it was the object of terrible persecution and torture. It may have been regarded as evil partly because of its alleged hatred of blue, the color "of the cloak of heaven" and that of the dress of the Virgin Mary. The cat was a striking figure in trials for witchcraft, was regarded as an imp of Satan, was accused of casting spells, and was girt about with mystery and superstitious fear.

In Flanders, cats were hurled from high towers on the second Wednesday of Lent. This custom persisted in Ypres until 1868 or later. In Picardy, cats were burned on the first Sunday of Lent. In Metz and other towns, they were sacrificed in bonfires on the evening of St. John. In England, they were hanged, burned by hundreds in mighty fires, roasted alive in brick ovens or at archery contests were tucked into leathern bottles and shot with arrows. In Scotland, they were impaled on spits and roasted alive before slow fires. From time to time on the continent, they were roasted in iron cages, over fires, in company with effigies of murderesses. The worrying of cats by dogs was a common sport. Boys tied cats together in pairs by their tails and hung them up to see them fight. Thus, persecution, fear and torment followed poor pussy through the ages until the eighteenth century, when superstition began to lose its hold. Even now, however, some terror of the cat remains in many lands; many persons regard her with aversion, if not with hatred, and so the old inheritance of fear still darkens pussy's pathway, and she keeps the attitude of apprehension as she slinks across the street.

The folklore of many peoples teems with superstitious cat tales and fables, many of them showing aversion, dispraise or suspicion. People still keep black cats away from the cradle in Germany.

Puss has a large place in literature and has added many words and proverbs to the languages of mankind. Fully fifty English words or phrases have been derived from her, and now in the twentieth century she is coming again to her own. Her star — eclipsed since the fall of the Goddess Pasht — again has reached its zenith. Carefully guarded from harm by humane societies, unrestrained by law or public sentiment, pampered, petted, worshipped almost as of old, "queen" of the cat show, attended by her most "humble slaves," puss faces the dawn of a new era. Dozens of books are devoted wholly or in part to chronicling the history, varieties, diseases, friends and enemies of cats, and everything pertaining to the beloved pet. There are cat magazines, cat clubs and cat homes. The attitude of present-day "cat worship"[1] is that the "queen" can do no wrong. A lady advertises in the "London Standard" for live birds with which to feed her cat. Another inserts the following notice in a Berlin paper: —

> Wanted, by a lady of rank, for adequate remuneration, a few well-behaved and respectably dressed children, to amuse a cat in delicate health two or three hours a day.

FITNESS, CHARACTER AND INTELLIGENCE.

The cat family (Felidæ) includes the lion, tiger, leopard, panther, cheetah, jaguar, ocelot, puma, lynx, ounce, wildcat and many small forms. There are at least sixty-six species scattered widely over the globe. This family always has been regarded by naturalists as carnivorous, rapacious, unsocial, cautious, some-

---

[1] This expression is not coined in derision, but is quoted from a cat lover's book.

times brave, sometimes cowardly with dangerous antagonists, but bold and courageous when brought to bay.

Naturalists agree that the cat is a highly organized and intelligent animal. Mivart says that no more complete example can be found of a perfectly organized living being. As compared with the dog, its intelligence is rated lower, but is probably underrated. The older naturalists assume that nature has destined animals of the genus *Felis* to subsist on the flesh of other animals. For this purpose she has endowed them with an "insatiably bloodthirsty disposition," and has furnished them with most effective means of destruction. Their exceedingly great strength, especially that of the jaw, their keen lacerating teeth, and strong, retractile claws, sharp-edged and pointed, are terribly efficacious in inflicting wounds, while their peculiarly flexible, agile bodies enable them to spring with great force upon their victims. All are regarded as exceedingly cruel, and the domestic cat as perhaps the most cruel of all, because of its habit of tormenting its prey.

Cruelty of the Cat.

Romanes says that the feelings which prompt a cat to torture a captured mouse are apparently delight in torturing for torture's sake. So far as he has been able to discover, the only other animals manifesting such feelings are man and the monkeys.[1] This cruelty, however, is not peculiar to *Felis domestica;* probably other small cats have similar habits. Foxes also have been known to "play" with their prey. Moreover, such habits cannot be considered blamable except in man, — the most viciously and knowingly cruel of living creatures. The cat evidently cannot realize as man can the poignant pains and terrible sufferings of its victims. Universally, the cat seems to take delight in torturing its prey, but this seems to be its means of developing the use  of its fore limbs, and it acquires a more perfect control over them than is possessed by any other domestic animal. By continually advancing and retreating, springing and striking, it develops the skill that enables it to pounce upon and strike down birds, insects and small mammals in flight, and to clutch its prey even in darkness. All the play of the kitten tends toward these ends.

---

[1] Romanes, George J.: Animal Intelligence, 1883, p. 413.

The Cat compared with the Dog.

In estimating the character and intelligence of the cat, it has been customary from time immemorial to compare it with the dog, much to the cat's detriment. The independence of the cat, its naturally unsocial character and its apparent lack of affection for its master place it in a very unfavorable light when compared with the sociability, affection and fidelity of the dog. Hamerton, who is evidently an admirer of cats, says that all who have written about them are of the opinion that their caressing ways bear reference chiefly to themselves; he says also that his cat loves the dog and horse exactly with the tender sentiments that we have for foot warmers and railway rugs during a journey in the depth of winter; nor has he been able to detect any worthier feeling towards himself. Continuing, he remarks that ladies often are fond of cats and pleasantly encourage the illusion that they are affectionate. Maiden ladies, he says, surround themselves with cats because of their inexhaustible kindness, and their love of neatness which is in harmony with the cat.[1]

Shaler, comparing the cat with the dog, shows that his experience corroborates that of the earlier naturalists. He says:—

> The cat is the creature of the domicile, caring more indeed for its dwelling place than it ever does for the inmates thereof. In a word, the creature must have come to us after our forefathers gave up the nomadic life. . . . Among the curious features connected with the association of the cat with man, we may note that it is the only animal which has been tolerated, esteemed, and at times worshipped, without having a single distinctly valuable quality. It is, in a small way, serviceable in keeping down the excessive development of small rodents, which from the beginning have been the self-invited guests of man. As it is in a certain indifferent way sympathetic, and by its caresses appears to indicate affection, it has awakened a measure of sympathy which it hardly deserves. I have been unable to find any authentic instances which go to show the existence in cats of any real love for their masters.[2]

Lest it may be said that Shaler's statement was inspired by antipathy, let me quote a few passages from cat lovers. Agnes Repplier says, in the introduction to a recent volume:—

> All nations have conspired to praise the animal which loves and serves. Few and cold are the praises given to the animal which seldom loves and never serves, which has only the grace of companionship to offer in place of the dog's passionate fidelity.[3]

---

[1] Hamerton, Philip Gilbert: Chapters on Animals, 1874, pp. 47, 48.
[2] Shaler, Nathaniel Southgate: Domesticated Animals, 1895, pp. 50, 51.
[3] Repplier, Agnes: The Cat, 1912, p. xiii.

Independence of the Cat.

Many cat lovers admire the cat because it loves not, because it is fond of the fire but not of the fire maker. Witness the following from Chateaubriand to M. de Marcellus: —

> I value in the cat the independent and almost ungrateful spirit which prevents her from attaching herself to any one, the indifference with which she passes from the salon to the housetop. When we caress her, she stretches herself and arches her back responsively; but this is because she feels an agreeable sensation, not because she takes a silly satisfaction, like the dog, in faithfully loving a thankless master. The cat lives alone, has no need of society, obeys only when she pleases, pretends to sleep that she may see the more clearly, and scratches everything on which she can lay her paw.[1]

The attitude of the cat toward man has been clearly stated by so many cat lovers that the facts may be regarded as established. The following, translated from "Un Peintre de Chats," by Henry Havard, states the case for the cat as he regards it: —

> This is the progress, and these are the admitted triumphs of the cat. She has conquered and domesticated man, reduced him to the rôle of an obedient servant, and required of him that he shall provide her with the luxuries she loves. In doing this, he but performs his duty, and need expect no gratitude. The loud declarations of naturalists count for little by the side of such a candid confession as that of M. de Cherville, who tells us in one of his charming essays that for two years he has obsequiously served a little cat, born under his roof, and raised by his careful hands. For two years he has studied her tastes, and shown her every attention in his power; and never in all this time has he won from her the smallest token of regard. Never has she vouchsafed him a caress by way of thanks, nor consented to go to him when called with loving words and tender cajoleries.[2]

Affections of the Cat.

Nevertheless, some psychologists claim to have found some evidences of real affection toward human beings in certain cats. Not all cats are alike. They vary as people vary, and abject slavery to a cat's every whim sometimes seems to win its real regard and affection, or at least its appreciation. Rarely is such service offered except by women, whose superlatively affectionate and maternal natures lead them to make any sacrifice for those they love, and sometimes to make even greater exertions to please when the object of their attentions manifests only indifference. Miss Winslow evidently had good reason to believe that her cat loved her. She says: "Do not tell me that cats never love people; that only places have real hold upon their

---

[1] Repplier, Agnes: The Cat, 1912, p. 9.   [2] *Ibid.*, pp. 62, 63.

affections. The Pretty Lady was contented wherever I, her most humble slave, went with her."[1] Many a puss has been known to be "contented" in the company of her "humble slave" and if such an attitude does not win the affections of a cat nothing will. There are many stories of cats which have refused food and died after the death of some human friend or benefactor, and such cats are always said to have died of grief, but, so far as I know, no post-mortem examination has been held in any such case to determine whether or not the cat died of disease.

It is well known that the female cat, like the females of other mammals manifests maternal affection, and that the male often murders his own offspring. It is well attested also that females when deprived of their young have been known to adopt those of other animals, and to suckle squirrels, rats, leverets, puppies, skunks, hedgehogs, and even to adopt young chickens, squabs and bobwhites. In the cases of mammals thus adopted the sucklings probably relieved the maternal fount and so gratified the cat, but the mothering of birds seems to be entirely altruistic. Cats like other animals have shown at times some evidence of attachment to domestic animals and even to birds, but such evidences of  affection are exceptional. Aside from such individual exceptions it seems to be accepted by the authorities as a fact that cats as a rule have a higher regard for the home than for its inmates. Shaler explains this in the following manner:—

> The differences as regards affection for localities which are shown by cats and dogs are perhaps to be accounted for by an original and essential variation in the habits of life in their wild ancestors. Judging by the kindred of the species which are known to us in their wild state, we may fairly suppose that the dogs were of old accustomed to range over a wide field, having no fixed place of abode; the pack ranging, if the occasion served, over hundreds of miles in any direction. On the other hand, with the cats, it is characteristic of the species that they have their lairs to which they resort, and a definite hunting ground on which they seek their food. They are, in a word, animals of a very determined routine. As there has been no effort by breeding to change this feature, it has remained in all its old ingrained intensity.

Most cats will return to their old home if possible rather than remain with the family at a new dwelling place. It is this trait of the cat's nature mainly that endangers the native wild life of woods and fields, as will be shown hereinafter.

---

[1] Winslow, Helen M.: Concerning Cats, My Own and Some Others, 1900, p. 9.

FECUNDITY OF THE CAT.

Cats are known to have from two to four broods yearly, with from five to nine in each brood. Fostered and protected from their enemies, a single pair might produce an enormous number in a few years. Hence the necessity for checking such increase promptly by killing all superfluous kittens soon after birth. An undue increase of the species must occur otherwise as cats have very few effective natural enemies in the New England States.

NATURAL ENEMIES OF THE CAT.

The domestic cat is preyed upon by the larger *Felidæ*, of which the puma and the two species of lynx are the only New England representatives. They are found rarely now except in remote and wild parts of the region. The *Canidæ* must be reckoned among the cat's natural enemies, but as the wolf is now practically extinct in New England, and as few dogs are bold and active enough to catch and kill cats, the fox is the only mammal which may endanger the domestic pet. Foxes have been seen to kill cats and carry them away from farmyards, and remains of cats sometimes have been found when fox burrows have been examined. On the other hand, a large, powerful cat has been seen to turn on a young fox and drive it away. Probably foxes make no serious inroads on the numbers of cats. Foxes, raccoons and even weasels may pick up a kitten in the woods occasionally, but it is improbable that any wild mammal appreciably reduces the numbers of cats in Massachusetts. The golden eagle preys on cats, but it is very rare in New England. I have known a great horned owl to attack and kill a full-grown cat at night, but never heard of another instance. The absence of effective natural enemies to check the increase of cats in New England goes far to explain the increase in the numbers of stray or feral cats roaming in field and forest. Man is the cat's best friend and also its greatest enemy, and it is in his power to control its numbers within reasonable bounds.

NUMBERS OF CATS.

In setting forth the effect of the feeding habits of the cat, it is essential first to give the reading public an adequate idea of the numbers and prevalence of cats, not only throughout cities, towns and villages of New England, but on farms and in forests as well, as no one who has not investigated the subject has any idea of their ubiquity. Hundreds roam about the country towns. On the early snows of winter their tracks may be found on nearly every farm in the land. There is no forest or woodland so remote that the cats have not penetrated. In 1912 I visited the Maine woods in December, and there, in the snow, miles from any human dwelling, were more tracks of cats than of any other creature.

Great Numbers of Vagrant Cats in Cities.

It is a well-known fact that cities are overrun by vagrant cats, many of them hungry and cold in winter, finding a precarious living by catching mice and rats and visiting "dumps" and garbage cans. Many are fleabitten, mangy and diseased, and the suffering among them must be great. All such cats should be executed, as a measure of humanity and public safety. Humane societies have undertaken this task in Boston, New York and other cities. The Animal Rescue League of Boston has done a great work in rescuing numbers of homeless, starving cats and humanely destroying them, also in disposing of surplus kittens. Mr. Huntington Smith, managing director of the league, has been kind enough to give me the following account of the cats handled by the association during ten years, and the disposition made of them: —

Year.	Received.	Destroyed.	Placed in Homes.
1905,	14,400	13,791	649
1906,	16,151	15,657	494
1907,	14,157	13,710	447
1908,	15,330	14,915	313
1909,	20,414	20,042	372
1910,	23,089	22,385	310
1911,	23,691	22,529	229
1912,	27,670	27,295	356
1913,	29,525	29,078	447
1914,	31,122	30,688	536
Aggregates, ten years, 1905–14,	215,449	210,090	2,908

It is noteworthy that in this time the number of cats destroyed annually increased more than 200 per cent. This seems to show an increasing multitude of cats annually bred in the city, but Mr. Smith explains this as follows: —

The increase in the number of cats taken by us is due, first, to a growing tendency on the part of the public in and around Boston to turn over to us animals that they cannot or do not wish to care for; to increased efficiency on our part by the establishment of receiving stations and an elaborate collection service; and to the fact that by the use of motor vehicles we are able to cover a much larger territory than ever before. These figures represent not only the city of Boston, but outlying towns and cities, more particularly Brookline, the Newtons, Cambridge, Somerville, Arlington, Everett, Malden, Revere and Chelsea. While the stray cat problem is still a serious one in the more densely populated part of this city, I think we are gradually getting it under control. On a single day two weeks ago we destroyed here at headquarters 269 cats and kittens.

Mr. Smith writes that it is a standing rule of the institution to give away only gelded male cats. Female cats are destroyed. In New York a similar necessary work is done by the American Society for the Prevention of Cruelty to Animals. President Wagstaff writes the editor of "Bird-Lore" that in 1900 the society put to death 257,403 cats, and in 1911, 303,949. Mr. Ernest Ingersoll of the National Association of Audubon Societies has kindly secured for me the following facts and figures regarding some more recent activities of this society: —

This society operates throughout the greater city, and picks up and humanely destroys "small animals" to the amounts recorded below: —

1911,	362,216
1912,	225,307
1913,	240,371
1914,	222,402

This includes dogs to the extent of about two-fifths or less.
This appears from the following particulars: —
"Small animals" have been destroyed during six months of the present year as follows: —

1915.	Cats.	Dogs.	1915.	Cats.	Dogs.
January,	10,774	3,567	April,	18,816	3,620
February,	10,887	2,620	May,	19,511	3,376
March,	16,417	3,382	June,	22,082	3,609

There were seized on the streets of this city in 1911, 50,956 cats; 1912, 24,624; 1913, 23,239; 1914, 22,265.

These figures should be greatly multiplied [writes Mr. Ingersoll] to get an idea of the total cats destroyed in those years, because many more are taken by request from houses than are picked up in the streets. An exception to this is the number for 1911, when a special series of night raids were made in the tenement district on the east side and 50,000 cats were caught. These night-wandering cats in the city are known as "ash-barrel" cats.

City cats make forays into the parks at night. A man employed to guard the birds in Central Park, New York, killed in six months, from January to June, 1910, 161 cats.[1]

If we consider the number of vagrant and superfluous cats in the city we well may wonder what the rate of increase may yet become in the country where cats, mainly nocturnal, may wander at will, unseen and unknown, and increase unchecked, except perhaps by the cold and starvation of winter, which generally they seem to survive.

Numbers of Vagabond or Wild House Cats in the Country.

Wild or feral house cats that pass their lives mainly in the fields or woods are seen rarely by human eyes, except by those of the hunter or naturalist. Therefore many people who have never investigated the matter, and never have seen such cats, find it hard to believe that they are numerous enough to be a great menace to wild life, but nearly all my most observant correspondents who roam the woods and fields report traces of many cats. Mr. William Brewster of Cambridge, the Nestor of New England ornithologists, says that he and his dogs frequently have started cats from their resting places in woods and game covers. He says, writing from Concord, that they are seldom noticed, being shy, elusive and largely nocturnal, but that he finds their tracks everywhere in the woods after the first snowfall. He asserts that his guides, James Bernier and William Sargent of Upton, Me., trappers of large experience, assured him some years ago that the forested parts of New England with which they were familiar were numerously inhabited by woods cats. Quite as many cats as other fur-bearing animals were caught in traps even in "locations upward of thirty miles from any house or clearing, and over the northern Maine line in the Canadian woods."

Mr. Charles E. Goodhue, naturalist of Penacook, N. H., says that it is hard to tell whether or not cats are vagrant or wild, but local trappers get many in their traps, and cats roam over the country in every direction. Three trappers among my correspondents corroborate this. Mr. Nathaniel Wentworth of Hudson, N. H., former game commissioner of that State, says that he has seen many cats sometimes miles away from any house, and feels sure that more game birds are killed by them than by the hunters, — an opinion expressed by very many others. Wm. C. Adams, Esq., a member of the Massachusetts Commission on

---

[1] Pearson, T. Gilbert: Bird-Lore, July-August, 1910, p. 174.

Fisheries and Game, has noticed particularly the tracks of cats in his travels. He found numerous cat tracks on the islands of Muskeget, Tuckernuck, Nantucket and Martha's Vineyard. On Nantucket he noted that the tracks extended three or four miles from any habitation. He saw traces of many birds evidently killed by cats, particularly on Muskeget and Martha's Vineyard. He describes a similar condition on Cape Cod, in the townships of Provincetown, Eastham, Orleans and Sandwich, where he has hunted. He says that cats are numerous in a large section between Worcester and the Rhode Island line, and in the country between Ware and Greenfield; also between Adams and North Adams, and in many parts of New Hampshire. He has observed many tracks on the winter snows; he has seen many cats, some of them with birds, and frequently has noticed them on lonely roads at night, by the light of his car lamps. Several hunters have told him of finding litters of kittens far back in the woods.

Mr. John B. Burnham, former chief game protector of New York, president of the American Game Protective and Propagation Association, says that his automobile lights frequently show cats at night. He has shot two recently more than a mile from any house and so heavily furred that they evidently were wild. Mr. Maunsell S. Crosby of Rhinebeck, N. Y. asserts that he killed fifteen on his farm in 1913, and he never molests any near the village, as they may belong to his neighbors. Mr. Lee S. Crandall, assistant curator of birds, New York Zoölogical Park, says that stray cats are numerous in that vicinity. Mr. Allan Keniston, deputy fish and game commissioner, Edgartown, writes that he has killed many wild or woods cats; has seen many tracks, and has seen cats kill meadowlarks and other birds. Mr. C. L. Gold, chairman of the bird committee of the Connecticut State Grange, at Cornwall, Conn., says that there are many there.

Mrs. Mabel Osgood Wright, Fairfield, Conn., president of the Connecticut Audubon Society, writes that in seven months, twenty-eight cats have been shot on her twenty acres, although the six nearest neighbors keep none. Mr. George C. Donaldson of Hamilton, member of the bird committee of the Massachusetts State Grange, avers that there are many cat tracks in the woods in that region. Hundreds of similar assertions might be printed would space allow, but a few abbreviated statements follow: —

"Hardly a day passes that I do not see one or more," Nathan W. Pratt, North Middleborough. "Saw at least twenty around a heronry, and judging from the tracks after a night's rain there must have been several times that number," Dr. C. L. Jones, Falmouth. "Have seen a great many cats in the woods and about abandoned farms and farm buildings that had not been occupied in many years, and far from any occupied building," C. Harry Morse, Belmont. "See many when shooting," Walter P. Henderson, Dover. "Have run across many in woods. Last year, killed three in one day far from any house," Samuel Hoar, Concord. "Legions of abandoned, vagrant, or wild cats," Bernard A. Bailey, M.D., Wiscasset, Me. "About one-half the tracks in the woods are cats' tracks," J. K. Jensen, Westwood. "In seven years I have destroyed thirty-five cats wandering in or near an extensive woodland area," William P. Wharton, Groton. "Often see wild cats in woods when hunting," Curtis Nye Smith, Newton. "Many seen on hills and marshes," Sarah E. Lakeman, Ipswich. "See plenty in the country when shooting," Vinton W. Mason, Cambridge. "Trap and kill about thirty per year, trying to get at chickens and pheasants," William Minot, Wareham. "Have seen many cats in woods. On any fresh snow, however far and thick the swamp, find cat tracks dogging those of rabbit and grouse, then signs of scuffle and feathers tell the tale," Clarence E. Richardson, Attleboro. "This fall and winter have seen about fifty to sixty," Harold K. Decker, West New Brighton, N.Y. "Over a dozen here," Hugh McCue, East Milton. "Constantly seen in the woods during the open season," E. Colfax Johnson, Shutesbury. "Tracks fairly abundant in the woods," G. B. Affleck, Springfield. "See a great many," Walter A. Larkin, Andover. "Many tracks can be seen after a light snow," Wm. B. Olney, Seekonk. "Neighbors have thanked me for killing fourteen in one summer," Julia W. Redfield, Pittsfield. "Too secretive to show themselves much, but their tracks are everywhere," Arthur C. Dyke, Bridgewater. "May be seen all over the woods, often shot by rabbit hunters," Thomas Graves, Plymouth.

The locations of these few reports, among many, show that the stray or feral cat is distributed widely. On the other hand, Mr. Hedley P. Carter of New Britain, Conn., says that he has hunted and fished for twenty-five years, and that he "scarcely ever sees a cat in the woods." Negative evidence, however, is of little value in the face of overwhelming positive evidence.

It is interesting to note the conditions under which this so-called domesticated animal has reverted to the wild state and spread over the country. It must be borne in mind that the cat, while partly tamed, has not been fully domesticated. It has not been subdued, confined or controlled, except in rare cases, but is to all intents and purposes a wild animal. In most cases it stays in the home of man, mainly because of the warmth of his fire, the food that it eats and its affection for the location where it was reared. If, by accident or design, anything occurs to interrupt its association with man, it readily returns to the wild. Shaler says: —

As a consequence of the affection which cats have for particular places, they often return to the wilderness when by chance the homes in which they have been reared are abandoned. Thus in New England, in those sections of the district where many farmsteads have of late years been deserted, the cats have remained about their ancient haunts and have become entirely wild. In this State they are bred in such numbers that their presence is now a serious menace to the birds and other weaker creatures of the country. The behavior of these feralized animals differs somewhat from that of creatures which have never been tamed. They have not the same immediate fear of man, but the least effort to approach them leads to their hasty flight.

Cats abandon Owners.

There are many other ways in which cats revert to a wild state. Cats are not all alike in disposition; occasionally one will leave its home and its master, walk out into the night and disappear, perhaps to return after months, perhaps never. Many leave good homes in the spring and take to the woods and fields, returning only when the approach of winter drives them to a nest in the haymow or to the master's fireside, but the most prolific cause of the return of cats to the feral state is not the fault of the animal, but that of man,— abandonment by their owners.

Owners abandon Cats.

Thousands of families go into the country or to the seaside in summer, taking cats or kittens with them, and leave their pets on their return to the city, not knowing, perhaps, that such cruelty is forbidden by law. Miss Winslow asserts that at Old Orchard Beach, Me., at the close of one summer, forty deserted cats were seen, and that sometimes as many as one hundred have been abandoned in a similar way at Nantasket Beach, near Boston. A report from Mr. Orrin C. Bourne, chief deputy fish and game commissioner of Massachusetts, asserts that one man killed thirteen cats that were deserted at Brant Rock at the end of the summer of 1914. Mr. Walter A. Larkin of Andover says that cats are left at summer camps in the woods when people leave them in the fall. He saw seven in one wooded tract in one day. Mr. Wm. H. Jones of Nantucket says that one hunter killed twenty-seven abandoned cats there last fall (1914). Many correspondents and people from all parts of New England report many cats abandoned by "summer people." Several persons note abandoned cats left uncared for in the city while their owners are away for the summer.

Many kindly people will not kill superfluous kittens, but cruelly leave them in the woods or by the wayside, in the hope, often a vain one, that some one will pick them up. One gentleman informs me that six were left at his door within a month; another that a kitten was left at his doorstep several times, but he refused to adopt it. Many such waifs either "go back to nature" or get their living from garbage cans, rubbish heaps, manure heaps and pigpens, killing whatever living things they can catch during the summer. Their tracks may be found on the first snows of winter as they wander, footsore and ravenous. A few of the weaker may succumb to storm and stress, but the hardy survive, to procreate their kind. This evil has gone so far that there is now no place where birds and game can be safe from this nocturnal enemy. Thirty-nine correspondents tell of people abandoning cats; 14 assert that they see many cat tracks on the snow; 46 that they often see stray cats in fields and woods; 51 that they see such in cities and towns, and 42 that they shoot them when known to be strays or seen far from houses in the woods.

It is difficult in many cases to determine whether or not cats are ownerless or merely astray from villages and cities. Cats continually radiate from centers of population. Many of them are homeless, others mere nocturnal wanderers, but most of them are destructive to bird life.

Cats unfed by Owners.

Many cats, never fed or half fed by their owners, forced to range in search of food, roam far at night. Mr. N. A. Nutt of South Ashburnham, whose work takes him out during the latter half of the night, has seen cats coming from a patch of woods on their way back to the village, across the railroad track, so wet with dew as to appear as if they had been plunged into water. Countless village cats, farm, stray and feral cats extend the rapacious influence of the species throughout the land. Dr. Frank M. Chapman of the American Museum of Natural History, New York City, believes that there are not less than 25,000,000 cats in the United States, and that there may be twice that number.[1]

---

[1] Bird-Lore, March-April, 1902, p. 70.

HABITS.

The following, quoted by Miss Repplier, as translated from the Latin by Thomas Berthelet and printed by Wynkyn de Worde in 1498, cannot be improved much to-day:—

The Cat is surely most like to the Leoparde, and hathe a great mouthe, and sharp teeth, and a long tongue, plyante, thin and subtle. He lappeth therewith when he drinketh, as other beastes do that have the nether lip shorter than the over; for, by cause of unevenness of lips, such beastes suck not in drinking, but lap and lick as Aristotle saith, and Plinius also. He is a swifte and merye beaste in youthe, and leapeth, and riseth on all things that are tofore him, and is led by a straw, and playeth therewith; and he is a righte heavye beaste in age, and full sleepye, and lyeth slyly in waite for Mice, and is ware where they bene more by smell than by sighte, and hunteth, and riseth on them in privy places. And when he taketh a Mouse, he playeth therewith, and eateth him after the play. He is a cruell beaste when he is wilde, and dwelleth in woods, and hunteth there small beastes as conies and hares.

The habits of the cat are so well known that comparatively little need be said about them here, but one error has been promulgated widely. The assertion that this animal can see in the dark is repeated by intelligent authors even to this day and should be corrected. No eye of flesh can see in absolute darkness. There must be some ray of light to render any vision possible. Undoubtedly, however, the cat and the owl can see much better in starlight or moonlight than we, but when cats catch mice or rats in dark cellars, where all light is shut out, it is because of the alertness of their senses of hearing, smell and touch. Rats and mice move about and live without inconvenience in utter darkness, and the cat, no doubt, is able to catch one now and then under the same conditions, but most of those that she catches probably are taken where there is a little light, in the dusk of morning or evening or in daylight.

The female cat naturally rears her young in holes in the ground, caves or hollow trees, from which she makes sallies over the country within a radius of a mile or more, striking down any animal which she can master and taking her kill to the den to provide for her young. She follows her prey into the tallest trees and into such dens and burrows as she can enter, but does not seem able to dig very well, and ever must lie in wait for the smaller burrowing animals. Much ink has been wasted in attempts to prove either that the cat was originally a native of treeless plains or that it belonged to a forested region. The probability is that it was derived from animals frequenting both plain and forest, but the tree is plainly its natural refuge of last resort. It is not sufficiently expert in climbing to follow the arboreal mammals with much chance of success, but it can reach their nests as well as those of birds, and being nocturnal it is able to attack many species on their nests at night.

FOOD.

The cat, being naturally carnivorous, feeds first of all on flesh, destroying birds, mammals, amphibians, reptiles, fishes, crustaceans and insects. Its path is a trail of blood. Nevertheless, it consumes some vegetation.

Vegetal Food of the Cat.

Cats naturally do not require much vegetable food, but they eat grass as a means of ridding their stomachs of indigestible portions of their food, such as the chitinous or shelly parts of insects, and bones, fur and feathers. The grass acts as an emetic when taken in small quantities and aids the stomach in regurgitating or throwing up indigestible materials. Hence the phrase "sick as a cat." Harrison Weir says that grass taken in large quantities acts as a purgative.

The species in domestication has become accustomed gradually to vegetable food, and a modification of the digestive system has occurred. The large intestine has grown longer and larger than in the wild cat, and thus the creature has become better fitted to digest vegetal aliment. Many domestic cats are fond of certain vegetables. Asparagus is eaten generally. Among the cooked vegetables eaten by individual cats may be named string beans, corn, potatoes, both cooked and raw, squash, pumpkin, beets, spinach and parsnips. Fruits, such as melons and olives, have been eaten in some cases, also chestnuts, cereals, macaroni, etc. Dog bread, white bread or corn bread often are fed to cats, with milk, meat juice or gravy. Some domestic cats will take almost anything that men eat, from peanuts to ice cream and candy, but others will accept little beside animal food.

Animal Food of the Cat.

No animals are disdained as food except such creatures as are protected by hard shells, spines or disagreeable scent or taste, and even these are killed whenever possible, even if they are not eaten. The cat, like man, the weasel, the peregrine falcon, and some other excessively rapacious creatures, often kills for pure lust of cruelty and slaughter, or for "sport," leaving its victims to lie where they fall. All the native or wild cats of

America, as well as those of other countries, are recognized as among the most destructive of all animals to game, birds and domestic animals, and therefore the policy of American communities has been for many years to offer bounties for the destruction of these animals as the best means to secure their extermination.

In considering the feeding habits of *Felis domestica*, the first striking and noteworthy fact that presents itself is that the hunting habits of the species are those of a solitary wild animal. It hunts alone, and will not be guided by human companions. Except in rare cases, it wanders at will, like any predatory wild beast, being as free from all human restraint or control as the lion, tiger, wolf or fox. Naturally nocturnal in habit, it hunts by night more than by day, thus largely concealing its depredations under the cloak of darkness.

The next important fact to be considered is that it has been introduced into America by man, to destroy other introduced species. It is not needed to maintain the biological balance established here for centuries, and, being released and allowed to run at large and increase with little check, naturally tends to disturb that balance, as all introduced forces may, with unfortunate results.

Having practically exterminated the wild cats of the eastern States, and having passed a national law forbidding the importation of noxious mammals and birds, we have in the meantime introduced another destructive species in vastly larger numbers and disseminated it throughout the land, so that it must live upon the country as the native cats formerly did, except that it has the advantage that, being considered a domesticated animal, it can go with impunity into places where native wild cats would be in danger. It can prowl around houses, gardens, poultry pens and orchards by day or night, where the fox, wolf or lynx would meet with a warm reception. Hence, because of its abundance, it has become more destructive to wild life about the dwellings of man than any other wild creature, and therefore more injurious or beneficial to man, according as it feeds, to a greater or less extent, on man's enemies or his friends.

Destruction of Insectivorous Birds by Cats.

The widespread dissemination of cats in the woods and in the open or farming country, and the destruction of birds by them, is a much more important matter than most people suspect, and is not to be lightly put aside, as it has an important bearing on the welfare of the human race.

The Cat a Birdcatcher in Ancient Times.

The ancients recognized the cat as a destroyer of birds. If we may judge from pictorial representations on the buildings,

The cat as a bird killer. (From an ancient Egyptian painting at Thebes.)

tombs and monuments of the ancient Egyptians, the principal early use made of the animal was as a killer and retriever of birds. To the ancient Egyptians, birds (except the sacred ibis and the hawk) meant just so much meat. Apparently these people were able to utilize the birdcatching propensities of the cat, and to train her even to enter the water and catch or retrieve waterfowl. In the Egyptian gallery of the British Museum there is a painting of a man in a boat engaged in throwing a crooked instrument like a boomerang at a flock of birds, and on the same tablet a cat much like our common, striped tabby,[1] but with longer legs and tail, is represented as seizing a duck by one wing while she has a short-tailed bird like a quail and another, apparently a songbird, under her feet. In such situations puss appears often on the monuments of the Middle Empire, but so far as I can learn she is not represented as catching mice or rats. Diodorus tells of a mountain in Numidia inhabited by a "commonwealth" of cats, so that no bird ever ventured to nest in its woods.

Cat strangling a bird. (From an ancient mosaic in the Neapolitan Museum.)

No remains of cats were found in Herculaneum or Pompeii, but in the museum at Naples are some mosaics that came from Pompeii which show that cats were known there, as they are represented as attacking or killing birds. Agathius, a writer of epigrams and a scholasticus at Constantinople, who lived from 527 to 565, in

---

[1] The word "tabby" does not refer to the sex of the cat but to its markings, which resemble those on watered silk, which was once known by the same name. See Harrison Weir in Our Cats and All about Them, 1889, p. 137.

the reign of Justinian, has left two epigrams in which he scores a cat for tearing off the head of a tame partridge.[1]

A poet of Bagdad bewails the fate of his cat killed with an arrow while robbing a dovecote, and Miss Repplier in one of her charming volumes reproduces his wail from the Arabic of Ibn Alalaf Alnaharwany;[2] but the most celebrated ancient poem bewailing the cat's destructive proclivities is the "Anathema Marantha" by John Skelton, in the "Boke of Phylyp Sparowe," in which he calls down upon the whole race of cats the vengeance of the gods, mankind and the monsters of all creation in punishment for the killing of a pet sparrow. The poem begins:—

Cat stalking birds at a fountain. (From an ancient mosaic in the Neapolitan Museum.)

> That vengeance I aske and crye
> By way of exclamacyon
> On all the whole nacyon
> Of cattes wild and tame
> God send them sorrowe and shame
> That cat especyally
> That slew so cruelly
> My lytell pretty sparrowe
> That I brought up at Carowe.

He devotes this cruel "catte" to the tender mercies of the lions, leopards, "dragones," the formidable "mantycors of the montaynes," and hopes that "the greedy gripes might tare out all thy trypes," and so on and on and on. The little bird's mistress also joins in the denunciation. She wails:—

> Those vylanous false cattes
> Were made for myse and rattes
> And not for byrdes smalle.

The Cat a Birdcatcher in Modern Times.

In every land, in every tongue, the cat has been noted as a slayer of birds. Maister Salmon, who published "The Compleat English Physician" in 1693, describes the cat as the mortal enemy of the rat, mouse "and every sort of bird which it seizes as its prey." The French and Germans, particularly, have deplored the destruction of birds by cats. M. Xavier Raspail, in an article on the protection of useful birds, written in 1894,

---

[1] The Cat, Past and Present, translated from the French of M. Champfleury (Jules François Félix Husson Fleury), with notes by Mrs. (Frances) Cashel Hoey, 1885, pp. 17, 18.
[2] Repplier, Agnes: The Cat, 1912, p. 42.

says that though cats are outside the law, and therefore may be killed with impunity, their numbers are renewed from the villages incessantly to such an extent that not a night passes without

traces of these "abominable marauders." Of 67 birds' nests observed from April to August, only 26 prospered; at least 15 certainly were destroyed by cats, and others may have been.[1] Baron Hans von Berlepsch, the first German authority on the protection of birds, after forty years' experience says that where birds are to be protected the domestic cat must not be allowed at large. The above are but a few citations, many of which might be made, to show that the cat always has been recognized as a menace to bird life. Many present day cat lovers, however, claim that their cats kill no birds, or very few, "not more than one or two a year," and that the destructiveness of the cat to-day has been exaggerated to the last degree. Hence, it will be necessary to give voluminous evidence of the bird-killing propensities of the animal. First, we will turn the pages of some of the volumes written by cat lovers. Harrison Weir avers that he was able to teach three cats not to kill birds that he fed about the door, but he never could break them of the habit of destroying many birds' nests.[2] The destruction of nests by cats at night usually is accompanied with that of the mother birds and the young. Sometimes only the eggs are ruined, but cats do not attack nests unless they are occupied.

Miss Helen Winslow says that her aunt in Greenfield had a cat that was in the habit of catching his own breakfast early each summer morning before the family was up, — a very common habit by the way. Invariably, she says, just before her aunt's rising hour the cat brought in a nice fat robin, unharmed, and penned it in the corner of the kitchen, apparently as a gift for the aunt. Although the bird always was set free the cat continued to catch one each morning *having first caught its own breakfast*. It would be interesting to know how many birds that cat ate that season beside those that it brought in. The remarkable assertion here is that the cat was able to produce a robin every morning, for it must not be supposed that it was able to catch the same robin many times in succession. One or two

---

[1] Bulletin de la Société Zoologique de France, Vol. 19, 1894, pp. 142-148.
[2] Weir, Harrison: Our Cats and All about Them, 1889, p. 15.

such experiences probably would be enough to drive a robin away from the neighborhood, or to render it too cautious to be caught again, but Miss Winslow says that for several summers the cat "kept up this practice." This tale illustrates the ability of the cat to catch birds.[1]

Birds cut by Claws of Cats may die.

It is probable that some of these robins died eventually from the blows of the cat's claws. It is not uncommon that a bird caught "apparently uninjured" is in reality fatally hurt by teeth or claws. In capturing so active a creature as a bird the cat must work quickly and savagely. Most of the birds thus taken are struck down by the extended claws, and since there are many authentic cases of so-called "blood poisoning" among human beings resulting from cat clawings and cat bites, some of which are said to have resulted fatally, in spite of medical attention (see page 86), many a bird which has been struck once by a cat, and released apparently uninjured, may suffer a lingering and agonizing death. Mr. Harry D. Eastman of Sherborn says that pigeons which have been cut by the claw of a cat usually "go light" and finally die, and that a gray squirrel caught by a cat, taken away at once and not bitten, refused to eat, and died a few days later.

Cat Poaching for Owner.

Gordon Stables seems to exult in the birdcatching habits of his pets. He uses the poaching habits of the cat to illustrate its devotion to its master by telling of a poor plowman who was ill. Meat was prescribed by the doctor, but the poor man was unable to buy it. Every day, however, until he recovered the cat brought him in a rabbit or a bird.[2] Miss Repplier tells of a lady near Belfast whose cat went poaching for her every day, thus providing her with partridges illegally, as she had no legal right to the possession of the birds;[3] but 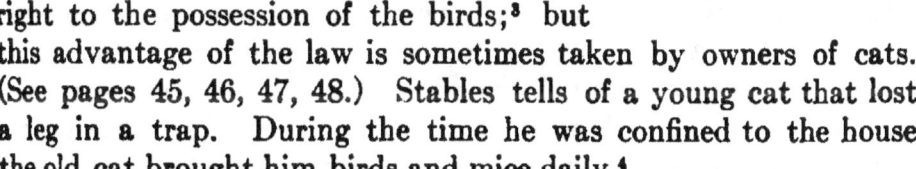 this advantage of the law is sometimes taken by owners of cats. (See pages 45, 46, 47, 48.) Stables tells of a young cat that lost a leg in a trap. During the time he was confined to the house the old cat brought him birds and mice daily.[4]

---

[1] Winslow, Helen M.: Concerning Cats, My Own and Some Others, 1900, p. 242.
[2] Stables, Gordon: The Domestic Cat, 1876, pp. 109, 110.
[3] Repplier, Agnes: The Fireside Sphinx, 1901, p. 242.
[4] Stables, Gordon: The Domestic Cat, 1876, pp. 111, 112.

Active and Intelligent Birdcatchers.

Again, Stables says that when Timby, a cat of which he knew, was but little more than a kitten he brought down birds from the highest trees.[1] He asserts also that he knew of a cat that caught two sparrows at once (probably young), and when pursued and attacked by a third sparrow (probably the mother) killed it with one paw.[2] This he considers "funny." Cats, he says, delight to spend a day in the woods, birdcatching. They rob the nests, too, when they find any, and cases have occurred of a cat paying visits to nests day after day until the young were hatched, then eating them.

Cats enticing Birds.

Romanes uses the birdcatching habit to illustrate the intelligence of the cat. He cites the statement made by Mr. James Hutchins (Nature, Vol. XII, p. 330), who says that a cat used as a decoy a young bird that had fallen out of a nest and made repeated attempts to catch the parents. He tells of a cat which often hid in the shrubbery and watched for birds whenever crumbs were thrown out; of another, having the same habit, that scattered crumbs for the birds that it might catch them when the family stopped feeding them; and of still another that, in order to attract the birds, uncovered the crumbs that had been covered with falling snow, and then crept behind a bush to await developments.[3] These stratagems met with varying success. Rev. J. G. Wood, a strong friend of pussy, avers that a cat concealed herself, decoyed sparrows within reach of her spring by imitating their note, and repeatedly caught them.[4] What chance would there be for a bird with cats so crafty? After all this, who, believing these tales, can doubt that cats are intelligent?

Numbers of Birds killed by Cats.

Most people do not realize how destructive cats are to bird life because their attention has never been called to the facts and because most feline depredations occur at night. In my investigations much evidence has been secured which is very convincing. In the year 1903, at the instance of the secretary of the State Board of Agriculture, an inquiry was undertaken regarding the decrease of birds in Massachusetts. As a part of this investigation a questionnaire was sent out to some 400 correspondents,

---

[1] Stables, Gordon: The Domestic Cat, 1876, p. 121.
[2] Ibid., p. 165.
[3] Romanes, George J.: Animal Intelligence, 1883, pp. 417, 418.
[4] Wood, J. G.: Natural History (1869), Vol. I., p. 201.

which was filled out and returned by more than 200. In response to a question regarding the effect produced on birds by their natural enemies, 82 correspondents reported cats as very destructive to birds. This was a much larger number than those reporting any other natural enemy as destructive. Nearly all who reported on the natural enemies of birds placed the cat first among destructive animals. These reports and opinions attracted my attention and I began to inquire regarding the numbers of birds killed by cats. The more the matter was investigated the more shocking it became.

Cats versus Spraying Trees. — Many people express the belief that most of the dead birds found have been poisoned by insecticides used in spraying trees. During three seasons, while investigating the effect produced upon birds by spraying trees, about sixty birds, adult and young, that had been picked up dead under or near trees sprayed with arsenate of lead, were sent me from various parts of the State. Each bird was skinned carefully, examined and dissected, and those which were not shown to have met death by violence were analyzed to see if poison could be found in them. Traces of lead and arsenic were found in two only. Others had met death in various ways, such as flying against wires or buildings; one had been shot; but nineteen showed marks of the teeth and claws of cats, and the coagulation of blood about the wounds showed that death had been caused by the attacks of cats. Evidently the cats were not hungry, but killed the birds in sport and let them lie. So far as this evidence goes, it indicates that cats are fully ten times more destructive to birds than is spraying as only birds killed by cats but not eaten could be accounted for.

Bird Slaughter by Cats. — Dr. Anne E. Perkins of Gowanda, N. Y., who has had a long experience with pets, tells of a cat which brought in meadowlarks, an oven-bird, two hummingbirds and a flicker within a few days.[1] She writes, "I am skeptical when any one says 'my cat never catches birds; it is only the hungry ones abandoned by their owners.' I have seen an active mother cat in one season devour the contents of almost every robin's nest in an orchard, even when tar, chicken wire and other preventatives were placed on the trunks of the trees. The robin builds so conspicuous and accessible a nest, and is so easily agitated by the approach of a cat, that it is difficult to save the young." She writes me that for years she has known of innumerable nests being robbed, those of robins, catbirds, song sparrows and wood thrushes especially, and she believes that the

---

[1] Bird-Lore, July–August, 1910, p. 174.

harm that cats do can hardly be overestimated. The young in the nests or just out most often fall a prey, but the cats caught many adult barn swallows, exterminated or drove away a colony of tree swallows, and caught snipe, grouse, hummingbirds, meadowlarks and many unidentified small birds. Many a time at 4 A.M. she has gone to the rescue of birds attacked by night-prowling cats.

Mrs. Elizabeth B. Davenport of Brattleboro, Vt., well known as an accurate observer, who has taken great pains to teach cats not to kill birds, writes that her experience covers many years while feeding birds about her grounds, and seasons spent on farms in Connecticut and in Vermont. In her grounds every small bird was attacked if cats had access to feeding places, and she had to surround these places with wire netting in summer and to protect them with high snow walls in winter. On the farm in summer cats brought in all kinds of ground-nesting or low-nesting birds. One cat in particular frequently brought in three or four birds a day.

Careful observers who have watched and protected birds for many years have had the best of opportunities for observing the destructiveness of cats. The editor of "Bird-Lore" publishes the statement from a correspondent that in one summer a neighbor's cat killed all the warblers on the place but one, eighteen in all, also two wrens, two woodpeckers and several other birds which were not identified.[1] Mrs. Oscar Oldburg of Chicago gives a partial list of birds killed by cats on her place, with dates. It contains fourteen individuals of six species and two nests full of eggs. She says also that many juncos are destroyed annually.[2]

Miss Cordelia J. Stanwood of Ellsworth, Me., says that at one time one of her neighbors kept seven cats. One of these in particular often caught as many as three birds a day, and is believed to have caught more when the young birds began to leave the nests. There were three cats in her own house, and her nephew who watched them said that they averaged more than three birds a day. She asserts that many persons in that region keep from three to seven cats, and she knows of one who keeps twenty. One day Mrs. Melville Smith, on whom she called, said that as she sat with a friend watching a hummingbird a cat caught it. The same day a cat kept at a house across the street caught four, and on the previous day a cat at the next house brought in two. The same day Miss Stanwood called on Mrs. Edward Wyman, and at her house the piazza was strewn with feathers of a black-throated green warbler. The number of cats

---

[1] Bird-Lore, January-February, 1909, p. 58.　　[2] *Ibid.*, July-August, 1910, p. 150.

kept in that family was from three to eight. They were well fed, but brought in birds ranging from warblers to woodcocks, and left them at the feet of members of the family. Two days later, when on her way to the home of a friend, she saw members of the family pursuing a kitten with a bird in its mouth. Within these few days another friend took her out driving, and related how a cat across the way had robbed a cedar waxwing's nest of five nestlings. She finds that since she has expressed an interest in the matter people, out of shame, conceal from her the depredations of their cats. That is a common experience. Miss Stanwood has a collection of bird skins, many of which were caught by cats. A naturalist whom I visited recently showed me a series of song sparrows' skins. Most of the birds had been killed by his two cats, which, he said, were continually catching birds. Many collections of this nature have been enriched by cats' victims.

Mr. Graham Forgie of Maynard, asserts that his cat kills about three birds daily. A lady recently informed me that her friend had a cat of which she was very proud because it was such a good hunter, and that in October it had killed and brought in twelve birds in two days. Nearly all these birds were myrtle warblers. Another lady reported last September that her cat, then having kittens, killed and brought in on an average two birds a day. During the fall migrations I have noticed that some cats kill more full-grown birds than at any other time. It is easy for cats to get them then for the following reasons: (1) Many of the birds then on their way south are the young of the year, that were reared in the great wilderness of the north, where there are few if any cats, and as these birds are young and inexperienced they do not realize the dangerous character of the animal. (2) The migrating sparrows feed mainly on the seeds of weeds at this season of the year, and so may be caught on or near the ground by the cat, which hides in the weed thickets. (3) On frosty mornings, warblers and thrushes find more insect food on or near the ground than higher in the trees, hence they come down in gardens and cultivated fields, where cats can easily hide and spring upon them. Those who feed birds on the ground in winter often attract them to places where they become the prey of cats, but the greatest toll is taken from the nestlings in spring and summer.

Young Birds the Chief Sufferers. — The young birds are either taken from the nest or caught on the ground before they have learned to fly well. Cats catch them with the greatest ease when they fall from the nest. Mr. A. M. Winslow of Worcester, says: "It is with sickening disgust that I recall the many species of birds, young and old, that were not only killed, but killed by slow torture, by cats on our place in the country. During the past five years in our yard in the city the robins have never succeeded in raising a brood of young ones which escaped the fate of being mauled to death by cats." Mr. F. H. Mosher of Melrose recently told me that robins had been very numerous in his neighborhood this year (1915), but that there were many cats roaming about the vicinity and that he believed that not one young robin escaped them; also, the killing of parent birds by cats leaves many young birds to starve in the nests.

I have observed some cases, and others have been reported to me, where cats have not noticed the young birds in the nests until they were nearly fledged, and then their cries for food apparently attracted the attention of their hereditary enemy, who, if watched and driven away in daylight, climbed the tree and got them at night. Dr. Robert T. Morris writes to the "New York Times" as follows of his two beautiful cats at the farm: —

It was observed that the cats would mark the location of each nest near the house by the calls of the young birds when they were being fed by their parents, and then would make the rounds of these nests every day, watching for the young when they struggled to the ground, as many young birds do in their first effort at flight. These two cats captured practically all the young from the nests of birds about the house, the number of young birds killed amounting to over fifty, to our knowledge, in the course of thirty days. The cats were then killed, although we were extremely fond of them as pets.

The following from J. O. Curtis, Mamaroneck, N. Y., July 24, 1914, explains itself: —

To the Editor of the New York Times: On Saturday last our cat caught two young robins. Having tasted blood, she has developed the hunting instinct, and during the last week has caught and killed seven birds. Her funeral will take place Sunday afternoon.

— Female cats with kittens often are very destructive to birds. I have known such a cat in June to destroy within twenty-four hours the young in six nests and also two of the parent birds, but this is the maximum, and I have never heard of another case so extreme except where cats have invaded dovecotes, chicken yards or pens in which birds were confined.

Much more detailed testimony is furnished by ornithologists and students of bird life. It is astonishing how rarely most people notice the cries of birds in distress, but the ornithologist recognizes them at once, and when he investigates he finds in a large proportion of the cases that the cat is the cause of the disturbance. No cat can kill so many birds in a season as can a bird-hawk, but probably there are two hundred cats in Massachusetts to every such hawk.

Mr. T. W. Burgess, editor for some years of "Good Housekeeping," states that although the dearest pet that he ever owned was a cat, he is beginning to see that the cherished pet is an agent more destructive than all others combined. He says that, one summer, weeks of watching and planning for photographs of birds at home came to naught through cats, as the nests of three pairs of robins, one of bluebirds, one of kingbirds and one of chipping sparrows in the orchard were emptied of their young by cats. Miss M. Purdon of Milton writes that she had her cat killed as the sight of countless birds and squirrels, half eaten or in process of being eaten, became too sickening to contemplate. The tragedies were so frequent that even the cook protested that they "made her feel sick." Mr. J. M. Van Huyck of Lee writes that he heard some robins screaming in the orchard, and when he rushed out four full-grown cats came out of the tree. They seemed to be strays, all after one robin's nest. Mr. Daniel Webster Spofford of Georgetown, writes as follows: "They watch the nests that they cannot climb up to, and when the young birds get so

All after one bird's nest.

they can tumble out of their nests, two or three cats stand ready to grab them, and run off with them, screaming, through the garden or street, and it is almost impossible to raise chickens or any kind of a bird without confining them in a close pen." Dr. C. H. Townsend, director of the New York Aquarium, writes from Greens Farms, Conn.: "Six nests of fledgling birds of various species were destroyed on our place last year by neighbors' cats, and they may have taken all there were."[1]

No one who has not witnessed the remarkable birdcatching feats of which a cat is capable has any idea of the imminence of this danger. My son, Lewis E. Forbush, last summer (1914) saw a large black cat approaching a young robin on the ground.

---

[1] Bird-Lore, July-August, 1913, p. 278.

He took the little bird and placed it on top of a wide, thick hedge nearly six feet high, believing that it would be safe; but the cat rushed, sprung, and vanished with the bird so quickly that it was hard to see how it was done, and it was all over before he could make a motion to interfere. Mr. Arthur W. Brockway writes from Hadlyme, Conn., that his mother, watching from the house, saw the family cat run up the pole of a martin box near by, seize a martin, and make off so quickly that she was unable to prevent it. Mr. Wilbur F. Smith, game warden of Fairfield County, Conn., says that when he was visiting one day in the country he found four cats tied in the yard, and was told that they were tethered there to keep them from catching birds. While the members of the family were at dinner, the young from a robin's nest fluttered to the ground, and the *tied cats caught them all*.[1] Birds often are taken from aviaries. Blackston tells how the cat gets them. He saw a cat apparently innocently watching the birds in his aviary, which he thought quite safe, as it was protected by zinc plates eighteen to twenty inches high. Suddenly the cat sprung and caught a fine singing canary, which had been clinging to the wires four feet or more from the ground, fastened her claws in the bird's body, and pulled it through the wires.[2] Cats sometimes kill penned game birds at night by reaching them through the wires. Several correspondents speak of seeing cats spring high and strike down birds in full flight, and they easily take slow-flying young birds in this way.

Statements from People in the Country. — In an attempt to get information regarding the comparative effectiveness of cats, traps and poisons in the destruction of rats, Mr. Walt F. McMahon visited 2 cities and 30 towns in 7 of the eastern counties in Massachusetts, in the months of August, September and October, 1914. Most of his work was done in a farming country, but he made many visits to villages. He secured 271 interviews from people who were willing to give information. Among them were the proprietors of 18 general stores, 5 livery stables and 8 grain stores. Inquiries were made also in regard to the number and kinds of birds caught by cats, but it was difficult to get this information because of recent agitation for a cat license. Many answers like the following were received: "Our cats do not catch birds, but Mrs. ——'s cats are catching them all the time;" or "Our cats don't kill birds. *We whip them if they do.*" Some owners admitted that their cats killed a few, while others asserted that their cats did not kill any. Still others

---

[1] Lacey, M. S. and L. A.: The Cat, What shall We do with it? p. 11. Published by the Audubon Society of the State of Connecticut.

[2] Blakston, W. A., and others: Canaries and Cage Birds (1880), p. 352.

believed that cats did not kill *many* birds. Some of the individual expressions are given below. Names are given only where special permission was granted. The names of towns are included to show the distribution of reports.

"You can't keep a cat from catching birds" (Lynnfield Center). "One bird a month" (South Sudbury). "Have never had a cat that would not catch birds. Don't think any nestlings got away" (South Hanover). "Most cats catch birds" (Hanson). "I never saw a smart cat that would not catch birds" (Hanson). "Cats catch one bird in two weeks" (Hanson). "You can't break a cat of catching birds once she gets a taste. Cats *will* catch them" (Sherborn). "Cats like better to catch birds than rats" (Sherborn). "Cat catches about one bird a week" (Billerica). "We raised one hundred and fifty chickens and the cats didn't touch one of them, so let them have the birds" (Littleton). "There are two or three nests in a tree near the house, and the cats get the young every year" (Hatchville). (A farmer of Danvers Highlands makes the same statement). "Had a cat that was something fierce on birds, killed forty-five chickens and brought in a half-grown pheasant" (Danvers Highlands). "This cat of ours will catch every bird she can get hold of" (Silas Hatch, Hatchville). "Robins and chipping sparrows nested here but no nestlings have been raised. Birds are scarce. Haven't seen a nestling robin this summer" (Eugene Hatch, Hatchville). "Cats make a business of catching birds" (James J. Hatch, Hatchville). "Catches all kinds of birds" (Hatchville).

Interviews with 271 people showed that the families or stores they represented kept 559 cats, 229 of which killed birds, according to the admissions of their owners (and more, according to their neighbors). Numbers of stray cats were reported in many cases, but the number could not usually be given exactly, as stray or feral cats cannot always be distinguished certainly from wandering neighborhood cats. Most people believe that stray cats are bird hunters.

Cats allowed to roam at Night. — The most significant item gathered from these reports is that out of 559 cats *405 are allowed to roam at night,*

A midnight marauder.

only 54 being kept in buildings. Many people who have studied the habits of the cat believe that the greatest numbers of birds are killed by it "between supper and breakfast," and unless the cat brings its game to the house, the owner has no knowledge of its nefarious work. Practically every cat that is allowed to roam at night where there are birds kills them sooner or later. As these 405 country cats were allowed to roam nightly where birds live, the chances are that every one of them caught a bird, adult or nestling, for breakfast time after time while its owner was still sleeping. Probably those 405 cats kill and eat thousands of birds yearly.

Correspondents report Many Birds killed. — The numbers of birds killed by cats cannot be approximated except by those who have paid particular attention to this subject. Among my correspondents are many such. Rev. Manley B. Townsend of Nashua, N. H., says that vagrant cats are common, and that nearly every day in the nesting season he has found birds killed and torn by cats. He has seen many fledglings in the possession of cats, and many reports of birds destroyed have come to him. Mr. Charles Crawford Gorst of Boston says that a friend told him that his cat had 14 birds laid out for its young one morning before breakfast. Mr. Samuel Hoar of Concord has known a cat to kill 10 birds in a day. Mr. H. Linwood White of Maynard tells me that a cat owned by one of his neighbors recently brought in 6 adult birds to her young in one day. Mr. Walter P. Henderson of Dover has seen a cat with 3 different birds in two hours. Mr. J. M. Van Huyck of Lee has seen cats hunting in the meadows for ground birds, getting both old and young, and striking down swallows as they flew over the grass. Mr. A. K. Learned of Gardner has known a cat to kill 9 tree swallows in one day. Mr. E. Colfax Johnson of Shutesbury says it is a common sight to see a cat eating a bird. Mr. D. T. Cowing of Russell asserts that his cat lived ten years and killed about 170 birds of which he knew, and believes that more were killed. Mr. Edward T. Hartman, secretary of the Massachusetts Civil Service League, says that where he lives he commonly sees cats hunting birds, and that he has known them to catch a great many. Mr. Frank E. Watson has no doubt that he has taken 100 birds away from his cat. Mr. George H. Hastings of Fitchburg had a cat that killed at least one bird a day in summer, and was known to kill 31 in one season. Mrs. Charles L. Goldthwait of Peabody called the attention of the owner of a cat to the fact that it had just killed a goldfinch; the owner said that the cat had killed several birds daily, and that it could not be prevented. Mr. A. M. Otterson of Hall, N. Y., has known a cat to kill 13 birds in a day, and to strike down swallows in flight. Mr. George G. Phillips, a member of the Bird Commission of Rhode Island, writes from Greene, R. I., that it is the commonest of sights to see cats hunting birds, and that the young in eight different nests about his house were destroyed by neighbors' cats last summer.

Mr. Frank Bruen of Bristol, Conn., writes that from the time robins come in the spring until they go in the fall there is an almost constant commotion, due to cats. He believes that half the young robins in the vicinity fall a prey to cats. Mr. R. L. Warner of Concord says that in his horseback riding about the country he constantly sees cats stalking birds, and frequently sees them eating birds. He often has seen cats climbing into trees to get at nests containing young robins. Mr. William Blanchard of Tyngsborough tells of seven robins' nests carefully watched and not one bird grew to maturity, all being devoured by cats. Mrs. Ella M. Beals of Marblehead tells of a farm cat with kittens which she watched, and which brought home several useful insect-eating birds every day and sometimes a few mice. Rev. Albert E. Hylan of Medfield says that he has known cats to bring in two or three birds a day for their kittens for some weeks at least. Mr. C. Emerson Brown, a Boston taxidermist, found the lair of two homeless cats. Near by was a heap of pieces of flying squirrels and red squirrels, and feathers of ruffed grouse and of many other kinds of birds. Dr. Loring W. Puffer of Brockton, now eighty-seven years old, and always an observer of nature, says that his experience shows that cats invariably will kill all the birds they can get. Mr. Nathan W. Pratt of Middleborough, frequently sees cats with birds. Mr. Samuel Buffington of Swansea has a cat that kills possibly one bird a day, and so many in the year that he has lost all account of the number. Mr. Sewall A. Faunce of Dorchester has known a cat to kill a bird "every morning" in summer.

Number of Birds killed per Day, Week, Month and Year. — Numerous correspondents have known individual cats to kill from 2 to 8 birds in a day, but the average is much smaller than this. Two hundred and twenty-six correspondents report the maximum number of birds they have known to be killed by 1 cat in a day, and the day's work for these 226 cats is 624 birds, or 2.7 birds per cat per day. Only 33 of my correspondents have kept any record of the number of birds killed by a cat in a week, but these 33 cats killed 239 birds in a week, or 7.9 birds per cat. Only 15 have kept any record of the number of birds killed in a month, and these 15 cats have killed 307 birds, or 20.4 birds per cat per month; but when we come to the record of the number of birds killed by a cat in a year, we find a different story. From 47 people we get reports showing that 47 cats have killed but 534 birds in a year. Evidently these are not the same cats that killed on an average 20.4 birds each a month. It is plain that many of those who have kept records of the cats that were killing large numbers of birds have either killed their cats before the year ended, which happened in several cases, or have failed to carry out their records for a year. Examination shows that most of the notes of a year's killing come from those who believe that their cats kill but few birds, and the notes are given casually, from memory. Some of these cats have been carefully watched, reproved, whipped, shut in or otherwise prevented from catching birds, while others are in city localities where they have little chance to kill birds. Still others are highbred, well-fed cats, which manifest little desire to catch anything.

The few people who have made continuous observations report that bird-killing cats in good hunting grounds, when not restrained, kill upwards of 50 birds per year. I have six such reports. It is not claimed, however, in any case that the cat did not kill more than 60, only that it was believed to have killed over 50. The most painstaking and careful report that I have was made by Mr. A. C. Dike. This has been recorded elsewhere.[1] The cat was a family pet. It was watched for one season and was known to kill 58 birds.

I have been widely misquoted as authority for the statement that every cat catches 50 birds per year, but my estimate was, that a mature cat in good hunting grounds will catch about 50 birds a year. Not all cats can or will do this. It would be impossible for any cat to kill such a number of birds where cats are numerous, for there would not be birds enough to "go 'round," nor would it be possible where birds are scarce, as in cities, where the birds available are largely house sparrows and doves which through centuries of association with men and cats have become hard to catch. Even in good hunting grounds only the most active bird-hunting cats can be depended upon to secure such a number of birds yearly, although no doubt some of them, particularly those that have run wild, kill many more.

Number of Birds killed in Various States. — My published statement, estimating the number of birds killed each year by the farm cats of Massachusetts alone, was given on the basis of 10 birds per cat per year, and 2 cats per farm. On this basis the farm cats of Massachusetts would kill about 700,000 birds

---

[1] Useful Birds and their Protection, 4th edition, 1913, p. 362 (Massachusetts State Board of Agriculture).

each year.[1] Through a typographical error, which was corrected in a later edition, the estimate allowed but 1 cat to a farm; but 2 was the figure used in the calculation, and our recent canvass seems to show that the farms average almost 3 cats each. The estimate has been deemed excessive by some, but has been regarded generally as conservative. Dr. George W. Field, chairman of the Massachusetts Commission on Fisheries and Game, estimates that there is at least 1 stray cat to every 100 acres in the State, and that each kills on the average at least 1 bird every ten days through the season, making the annual destruction of birds by stray cats in the State approximate 2,000,000. Dr. A. K. Fisher, in charge of Economic Investigations of the Biological Survey, estimates that the cats of New York State destroy 3,500,000 birds annually. Mr. Albert H. Pratt calculates that the farm cats of Illinois kill 2,508,530 birds yearly. Various estimates have been made concerning the number of birds killed annually by cats in New England. They vary from 500,000 to 5,000,000. Considering the above figures my own seem fairly conservative.

Destruction of Game Birds by Cats.

Perhaps the game bird most commonly killed by the cat in southern New England is the bobwhite. This species, one of the most useful of all birds to the farmer, highly valued as a game bird, frequents grass fields, gardens, grain fields, and weed and bush thickets where the cat hunts. Sportsmen say that they very often find cats in "quail covers," and not infrequently see them with the birds in their mouths.

Bobwhites.

Mr. Fred A. Olds saw a cat spring into the air and come down with a full-grown cock bobwhite in its claws.[2] Col. Charles E. Johnson asserts that he saw a cat with a bobwhite in its mouth running toward a negro cabin. When the colonel arrived at the cabin he found a colored woman plucking the bird. She said that the cat brought in birds very often.[2] Many cats are encouraged by their owners to bring in game. T. B. Johnson says in "The Vermin Destroyer," that he has known several cats that caught game and brought it home. These cats were highly esteemed by their owners.[3] (See also pages 33, 46–48.)

---

[1] Useful Birds and their Protection, 1907, p. 363 (Massachusetts State Board of Agriculture).
[2] Forest and Stream, July 29, 1911, Vol. 77, p. 175.
[3] Johnson, T. B.: The Vermin Destroyer, Liverpool, 1831, p. 27.

Mr. F. W. Henderson tells in the Rockland "Independent" of a cat that brought her kittens an entire brood of bobwhites. Dr. George W. Field, chairman of the Massachusetts Commission on Fisheries and Game, relates that a covey of bobwhites which he was watching in Sharon, was discovered by a cat and attacked at night, at intervals of two to seven days, until the number had become reduced from 16 to 8. They then left in a body for Canton, where they were recognized later. Mr. E. Colfax Johnson of Shutesbury says that he has known of entire flocks of young bobwhites being destroyed by cats. Mr. John M. Crampton, superintendent for the Connecticut State Board of Fisheries and Game, writes that last fall (1914) a farmer requested that a special protector be sent to look after the bobwhites on his land. When the warden arrived he found that the farmer had 15 cats, some of which had brought in 3 bobwhites already that morning. Mr. B. S. Blake of Webster tells of a cat that took home 3 bobwhites in one week. Mr. Edward L. Parker tells of a servant who saw a cat break up 2 bobwhites' nests. Senator Louis Hilsendegen of Michigan asserts in the "Sportsmen's Review" that Henry Ford bought 200 pairs of bobwhites at $3 a pair, and released them on his farm at Dearborn, Mich. A stray cat, left by a farmer who had moved away, found them, and it was noticed that their numbers were decreasing rapidly. A watch was set for the cat; it was shot and found to weigh sixteen pounds. Under a rail shelter, where the birds had fed, a mass of feathers and other remains about a foot deep was found. That cat, says the senator, had killed more than 200 bobwhites which had cost the owner over $300. Mr. E. R. Bryant of the Henry Ford farms writes me that this story is true except that it may be a little overdrawn in regard to the number of birds killed. He never knew exactly how many were slain by this cat.

Ruffed Grouse.

Cats are nearly as destructive to grouse as to bobwhites. I have seen a ruffed grouse that was killed on her nest and partly eaten by a cat, while the eggs were scattered and some were broken, but not eaten. Almost invariably in such cases a careful search will reveal a few hairs of the cat on some branch or twig, lost in the struggle. If several steel traps be set carefully concealed around the dead bird the cat may be taken.

Mr. William Brewster tells of a day's hunt by four sportsmen with their dogs, in which they killed but one game bird — a bobwhite. On their return at night to the farmhouse where they were staying they found that the farm cat had beaten their score, having brought in during the day two bobwhites and one grouse. Mr. Cassius Tirrell of South Weymouth asserts that a cat living not far from his home has brought in so many bobwhites and grouse that the family has "lost track of the number." Mr. John B. Burnham of New York, president of the American Game Protective and Propagation Society, writes that one of his farmer's cats killed "quite a number" of ruffed grouse, including adult birds. Several correspondents report cats seen carrying or eating full-grown ruffed grouse, and one saw a cat catching the young. The illustration of the dead grouse presented herewith is that of a bird killed Feb. 2, 1915, by a cat which was frightened away while in the act. The bird was not quite dead, but its throat was torn open and it was breathing its last. (See Plate VI.)

Heath Hens.

Probably the cat is, next to man, a chief factor in the destruction of the prairie chicken on the plains. Miss Althea R. Sherman writes me from National, Ia., that the farmers there keep from 12 to 18 cats per farm, and that she does not know of one that will not hunt birds. The prairie chicken is much like the heath hen, which has been almost exterminated in the east. The cat and the rat are the only predatory mammals on Martha's Vineyard, where the few remaining heath hens now live, and whenever cats come on the reservation, the remains of full-grown heath hens tell the tale. Therefore, Superintendent Day kills every cat of which he finds traces. (See Plate II.)

Pheasants and Partridges.

Since the introduced ring-necked pheasant has become common in Massachusetts, many reports of the killing of these birds by cats have been received. They are taken from the time the chicks are hatched until they are full-grown, although the young birds and females suffer most. I have seen two full-grown cocks that had been killed by cats, and many more have been reported. This seems remarkable, as the cock pheasant is said to be a great fighter and to be able to whip the ordinary barnyard cock. Mr. Lee S. Crandall, of the New York Zoölogical Park, writes that he has known of several instances where cats have killed and carried off full-grown golden pheasants, and that they have killed so-called Hungarian partridges in the park. It is a well-known fact that many of these partridges, imported into Connecticut by the game commissioners at an expense of many thousands of dollars, were killed by cats. Some cats specialize particularly on certain game birds.

Snipe, Woodcock and Other Game Birds.

According to Darwin, a Mr. St. John records a case where a cat frequented marshy ground at night and brought home snipe and woodcock. Mr. W. F. Henderson of Rockland tells of a man whose cat brought in 18 woodcock in a season. Rails are common game of cats. Prof. Edward P. St. John of Hartford, Conn., tells of 12 Virginia rails brought in by one cat. All the shore birds, plover and snipe, are taken by cats, particularly the young of those that breed in inhabited regions. No species of game bird, except possibly certain wildfowl, can escape the toll that cats take of their numbers. This tax is severe enough with wild birds breeding naturally, but when any attempt is made to rear large numbers of game birds on a small area, as on a game preserve or bird reservation, the cats' destructiveness is multiplied tremendously.

The Cat on the Game Preserve.

All experienced gamekeepers regard this animal as one of the most vicious and despicable of the so-called vermin which often render the raising of game birds a precarious calling. Prof. Clifton F. Hodge, a pioneer in the successful artificial rearing of grouse and bobwhites, was almost forced by cats to give up his experiments in Worcester. Although the birds were kept in pens, the cats reached through the wires at night, tore, mutilated and killed many birds, and drove the brooding mothers from their young, so that the little ones died of exposure; and when, with the utmost care and vigilance, bobwhites were reared and liberated, the cats caught practically all in the fields. The remarks of gamekeepers about cats' ravages are unprintable, and they rarely attempt to rear game birds without first destroying all roaming cats if possible.

I have followed the history of several undertakings of this character. In one instance the keeper on a game farm fully one mile from any village, and with very few neighbors, was obliged to destroy about 200 cats the first year, as the cats got all the young birds. In two other cases nearly half that number of cats were destroyed. On the Childs-Walcott Preserve, in Norfolk, Conn., which is situated in a rather wild, mountainous country, 81 cats were taken from February, 1911, to September, 1913.[1]

---

[1] Job, Herbert K.: The Propagation of Wild Birds, 1915, p. 5.

Mr. W. R. Bryant, of the Henry Ford farms, Dearborn, Mich., says that it has been necessary there, in protecting birds, to kill " about 75 cats each year, or possibly less each succeeding year." He names the house cat as the first and greatest drawback "in our efforts to save and increase the song birds and game birds." Such destruction of cats is a necessity; otherwise practically no game can be raised. Mr. Harry T. Rogers, of the New York State game farm, tried for some time to kill 5 cats that invaded the premises in 1914. These cats became so troublesome that an organized hunt was made for them, but Mr. Herbert K. Job asserts that before they were killed their depredations had cost the State of New York fully $1,000.

Number of Observers reporting Game Birds killed.

Forty-six observers write me that they have known cats to catch and kill ruffed grouse; 44 report the same of bobwhites; 12 report pheasants; 11, woodcock; 8, rails; 3, heath hens; 3, shore birds; 2, mourning doves; and 2, wild ducks.

Destruction of Poultry and Pigeons by Cats.

Every one knows that some cats kill chickens and that such cats usually are short lived, as the owner of the chickens commonly requisitions the shotgun as soon as he is aware of the identity of the marauder. He often will allow his cat to kill song birds to its heart's content, but chicken killing is quite another matter. Nevertheless, if we accept the statements of my 400 correspondents as indicative of the general situation, more chicks than birds are known to be killed by cats. This is readily explained, for *no one ever knows how many birds a cat kills if it is allowed to roam*, while chicks are counted and watched, and the numbers killed by cats can be approximated closely.

Chickens.

Mr. Charles M. Field of Shrewsbury has known a cat to kill 18 chicks in a day. Mr. Frederick W. Goodwin of East Boston gives a record of 24 killed by a cat in one day. Miss Mabel McRae, Boylston, has a record of 25. Mr. A. B. Brundage of Danbury, Conn., tells of 34 as a day's work for one lusty cat. Mr. Wilbur F. Smith of South Norwalk, Conn., says that one of his neighbors lost over 40 chicks before he began to shoot. He got four cats and the chick killing ended. Mr. J. Riley Rogers of Byfield writes that he knows of one cat that got 60 in one night. This evidently was due to carelessness in leaving doors open at night. The ordinary chicken killer gets from 2 or 3 to 12 in a day, and usually its career is short, except where the chickens wander into shrubbery or woods, where the cat can creep on them unseen by the owner. In such cases the losses are serious and long continued. I have lost many chickens by cats in this way.

Mr. Warren H. Manning of Boston has known a cat to kill between 60 and 90 chickens in a week. Mr. William H. Learned of East Foxborough has known one to kill 64 within a month. Mr. Clayton E. Stone of Lunenburg says that one of his neighbors lost over 75 in one season, and that one stray tomcat destroyed over 100 chickens in his neighborhood in one summer, some of which were nearly half grown.

Mr. E. G. Russell of Lynnfield says that he has killed 14 cats that stole chicks. Many people keeping from 1 to 4 cats each report the killing of from 20 to 75 chicks in a season by rats that the cats failed to catch.

It is of interest to examine the figures from reports regarding the number of chickens killed by cats; 124 cats killed 685 chickens in one day, or 5.6 chickens each. The number reported as killing chickens for a week is much smaller, as many chicken killers are not allowed to live a week after their misdeeds become known. Twenty-four cats killed 396 chickens in a week, an average of 16.5 chicks per cat; 11 cats killed 189 chicks in a month, or 18.8 per cat, and 18 cats killed 699 chicks in a season, or 38.8 each. The last were mostly vagrant or woods cats which took chicks, notwithstanding the efforts of the owners of the chicks to stop it. The above is a remarkable showing when it is considered that strenuous efforts were made to stop these depredations, and that nearly all these cats were killed within a short time after it became known that they were killing chickens.

Most of the chicks killed by these sporting felines are small, but it is not rare for them to attack chickens from one to two pounds in weight, or even larger. Farm cats do not commonly attack chickens, owing to early education and the quick elimination of the chicken-killing strains, but the city and village pussies, and stray or feral cats not subject to this early training and later selection, furnish most of the chicken killers. Mr. Sewall A. Faunce of Boston says that his cat caught a half-grown rooster, brought it home and was killing it when he came to the rescue. Mr. Newell A. Eddy of Bay City, Mich., missed chickens day after day from a flock about one-third grown, and finally his hired man discovered that they went away in company with a large cat. Mr. F. C. Stevens of Somerville tells of a kitten owned by Mr. John Little of Salisbury, N. H., that appeared to be playing with half-grown chickens. It killed one and then another. *Exit kitten!* Mr. Little had similar experiences with other cats. Mr. Philip Laurent of Philadelphia asserts that a black male cat was accustomed to sleep all day in his yard, prowling at night, and on several occasions he saw the cat in the yard early in the morning with chickens, weighing from two to three pounds each, which it had killed. Dr. Louis B. Bishop, the well-known ornithologist of New Haven, Conn., writes that in October or November a gardener employed by one of his neighbors said that cats had killed two chickens and left the remains in the yard. Dr. Bishop did not see these chickens, but from the date believed them to be nearly, if not quite, full grown. The gardener believed them to be spring chickens, about six months old.

Young Turkeys.

Mr. Richard H. Barlow, president of the Lawrence Natural History Society, avers that when he was with his uncle, Samuel Benson, at Manchester, Eng., about 1873, they had a half-grown black and white kitten that was turned out to shift for itself. It disappeared for nearly a year. Then they began to miss young turkeys from valuable prize stock, from the size of quails up to three pounds in weight. After about 40 had been lost, a trap was set, baited with a young turkey, and an immense cat was caught weighing 17½ pounds[1] and marked exactly like the lost kitten. Mr. Barlow is not sure how much of the weight was cat and how much turkey, but no more turkeys disappeared. Any cat that will catch large chickens and young turkeys is likely to kill small full-grown fowls.

Bantam Fowls.

Mr. Ross Vardon of Greenwood says that his cat caught a full-grown bantam which she dropped when chased, but it died. Mr. A. K. Learned of Gardner says that eight or nine years ago his cat went to Mr. James Hemenway's place, some thirty rods away, killed a bantam hen and brought it home. The cat's career was cut short. Mr. James M. Pulley of Melrose, says that about Dec. 27, 1914, he saw a black cat run crouching among his bantams, pick up a two and one-half year old hen and carry it off. He asserts that he has lost about a dozen, presumably in

---

[1] Harrison Weir has recorded a cat weighing 23 pounds. Other records exceed his.

the same way, as his neighbors have seen cats carrying off his fowls. Some of these were "half-breed" bantams about as large as a Leghorn hen. Previous to this occurrence he chloroformed a cat that took several nearly full-grown Minorcas from the premises of a near neighbor. It is but a step from such work as this to the killing of full-grown fowls of standard breeds.

Full-sized Fowls.

The number of reports received regarding the killing of full-grown domestic fowls by cats is surprising, but it is well known that some of the wild species from which our domestic cat probably was derived are destructive to poultry, and some house cats which run wild revert to these original habits. I have not found much evidence in cat literature regarding the destruction of standard sized fowls, but Finn remarks that crossbreeds between long-haired and short-haired cats are likely to become poachers, and will even attack full-grown fowls, which, he says, is a rare fault of ordinary cats, although fowls are an important part of the natural food of wild cats.[1] "Forest and Stream" says that in South Africa farmers suffer much from the numerous wild cats, which are very destructive to lambs, kids and fowls. The progeny of domestic cats often run wild and are most dreaded as having more than the usual amount of cunning.[2]

Miss Repplier asserts that the cat is described in ancient documents as a hunter of mice and a slayer of hens,[3] and the evidence submitted below seems conclusive that the latter habit, though uncommon, still persists.

Having lost fourteen hens by a supposed dog or fox, I had the fowls shut in. About November 1, a fine, white Plymouth Rock pullet, nearly full grown, was found in the henyard partly eaten. It did not seem probable that any dog or fox could get over the high wire fence, and the appearance of the carcass was similar to that of a grouse killed by a cat. It is well known that cats, from the lion and tiger down to the household pet, are almost certain to come back at night to their partly eaten prey, and may be shot or trapped then. Three traps were set, and that night the largest cat in my experience was caught. No more fowls were taken or killed. There is much more circumstantial evidence that points to the cat as a destroyer of grown poultry. Mr. Thomas Aspinwall of Brookline shot several cats that at different times stalked his father's hens with the apparent intention

---

[1] Finn, Frank: Pets and How to keep Them, 1907, p. 18.
[2] Forest and Stream, Nov. 1, 1902, Vol. 59, p. 345.
[3] Repplier, Agnes: The Fireside Sphinx, 1901, p. 11.

of attacking them. Mr. A. W. Streeter of Winchendon asserts that a hen that was beheaded and left to bleed was pounced on by a cat, dragged off and partly eaten before it was found, half an hour later. Mr. Daniel W. Deane of Fairhaven says that he never knew a cat with a good home to kill a full-grown fowl, but whenever in his long life he has found a hen killed and partly eaten, he has surrounded the carcass with traps, and almost invariably got a cat the next morning, and sometimes two. Lest it may be objected that circumstantial evidence is not conclusive the testimony of eye witnesses must be given.

Mr. Charles W. Prescott, a resident of Concord, reports that he lost a large fowl that was taken out of his henhouse window, which was 5 feet 6 inches from the ground. He tracked the animal 400 yards, found the fowl partly eaten, took it back to the henyard, lay in wait that night, and shot a large yellow cat when it appeared and started to drag its prey away. He said that the cat weighed almost 20 pounds. Mrs. Cora E. Pease of Malden tells of a large, cream colored Angora cat named Richard Mansfield that brought home fowls to its mistress in 1901 from a neighboring poultry yard, but so far as she is aware the birds were not seriously injured and were released by the cat's owner. Richard was a very high-bred cat and would eat little but cream and beefsteak, according to his owner. Evidently the hens were taken in sport.

Mr. Franklin P. Shumway of Melrose saw a cat spring on and kill a hen that had stolen away and made a nest in the under-

The fowl killer.

brush. This occurred at his country place in Forestdale about May, 1912. Mr. Freeman B. Currier of Newburyport tells of a cat kept in the family of Mr. James P. O'Neil which had the habit of chasing hens out of the yard, in which sport it was encouraged by its owner. Soon it began to kill them, and no one was able to stop it. Mr. Geo. W. Piper of Andover relates that he heard a hen squawking when he came home one night at 9 o'clock. He went into the barnyard and saw a cat killing a hen. The next night he lay in wait for it and shot it as it came back. Mr. Harold K. Decker of West New Brighton, N. Y., says that two hens were killed at night and several others wounded by a cat belonging to Mr. C. M. Smith of Westerleigh. This cat got into the coop at one of the small doors, which had been inadvertently left open. Once a tomcat owned by a neighbor got in through Mr. Decker's henhouse window, attacked a cock, tore out much of his plumage, and mangled the bird severely, but the noise of the struggle roused the household and Mr. Decker got out in time to save the rooster. Miss Agnes C. Eames of Wilmington says that a townsman saw his cat leap upon one of his own hens, seize it by the back of the neck and kill it. It was given no opportunity to kill another. Mr. L. H. Howe of Newton tells of a cat that killed a hen and brought it home. Mr. Clarence E. Richardson of Attleboro, while trapping, came upon a cat eating a full-grown fowl, freshly killed. When it saw him, it started to carry off the hen, but he interrupted the proceeding at that point. Mr. William Dutcher of Plainfield, N. J., president of the National Association of Audubon Societies, says that he has known a cat to kill a full-grown fowl, and Mr. Albert E. Shedd of Sharon says that a friend reported the killing of a large Brahma fowl by a 15-pound cat in Providence, R. I. Mr. Perkins R. Livermore of Marshfield Hills writes: "Some years ago I had a henhouse up back on my place near the woods. I found that something was killing my hens. I set a steel trap and caught a big woods cat. He had killed fifteen hens during a period of two or three weeks." The catching of the cat ended the killing of the fowls. If the above statements from reputable witnesses approximate the facts, the larger vagrant or woods cat may yet become as great a menace to the poultry industry as the fox. Possibly many cases where fox or skunk have been blamed might have been traced to the cat. Cats are large and strong enough to kill full-grown fowls with ease. The larger cats are much heavier than the ordinary fox, and it is well known that skunks, minks, weasels and even rats have killed many fowls at night.

It is only just to the cat to say that many cats which catch rats, but not chickens, are very useful in destroying rats about henhouses, and that rats are sometimes fully as destructive to chickens as are untrained cats.

Pigeons or Doves.

There are many complaints regarding the killing of doves by cats. Twenty-four correspondents report this. It would seem difficult for a cat to catch so watchful a bird as a dove in the open, but a practiced dove killer does not need to steal up very near to endanger its victim. When the experienced cat has crept within the proper distance it catches the dove in two bounds. The first does not bring it within striking distance, but with the second it often reaches the dove, already in the air, and strikes it down with its forepaws. Some cats become very expert at this game. Cats often miss their prey, but this is true even of the swiftest hawk.

Prof. John Robinson of Salem writes that a flock of pigeons has been homing in the barn of the Robinson family for eighty years, and that it has been necessary to keep up a persistent and unceasing fight to protect them from cats. About twenty-five years ago in the battle with the cats, 25 were killed in one year, 30 in another, and about 20 more in some succeeding years; after that cats were killed only as special marauders became intolerable. Pigeon breeders complain that, even when their birds are confined in wire netting enclosures, cats spring upon the wire by day or night, and, reaching through, tear the birds. Occasionally a killer finds its way into a pigeon loft at night, and nearly wipes out the flock. Mr. William D. Corliss of Gloucester says that about thirty years ago a house cat owned by a Mr. Lowe got into the dovecote of William Corliss at night and killed about thirty fancy pigeons, — pouters, fantails, etc. Members of the family say that this cat did not attempt to eat the birds but tore open their throats and is believed to have drunk the blood. Mr. Harry D. Eastman of Sherborn had a large flock of fancy pigeons, but the neighbors' cats killed "over one hundred dollars worth," and he gave up keeping them.

Cats eating Eggs.

Harrison Weir seems to believe that cats commonly eat birds' eggs in England, but I have never known a Massachusetts cat to eat an egg. Sometimes the eggs in a nest are broken when the mother bird is caught by a cat, but usually they are not eaten, and this has always seemed characteristic of attacks by cats. Nevertheless, in my reading, several instances were noted where cats were seen to eat birds' eggs or hens' eggs. A cat in a grocery learned to roll eggs to the floor that they might be broken for her repast, but this habit is exceptional. Mrs. Margaret Morse Nice tells of cats in Oklahoma becoming a great nuisance by breaking and eating hens' eggs.

EXTERMINATION OF ISLAND BIRDS BY CATS.

An isolated island is a little world by itself, and any fertile, well-watered one where birds can be protected from their natural enemies is likely to become a bird paradise. Gardiner's Island, N. Y., has been noted for many years for the numbers of birds that breed there, and for their tameness, although gunning is allowed upon the island during the shooting season. There are no cats there.[1] Wherever cats have been introduced and allowed to multiply unchecked upon an island, they have decimated, driven out or exterminated the birds.

Rothschild, in his great work, "Extinct Birds," names the cat first after man among the only important exterminative agents, and gives instances of the extermination of birds on sea islands. Henry Travis, the New Zealand ornithologist, says that many of the islands in that part of the world formerly teeming with bird life are now denuded because of the introduction of the cat. On the Chatham Islands, five hundred miles east of New Zealand, a land rail, *Cabalus dieffenbachi*, and a long-tailed wren-like bird, *Bowlderia rufescens*, are now believed to be extinct. Another land rail, *Cabalus modestes*, on the Island of Mangare, formerly found also on Warekauri, has become extinct since the invasion of cats.[2] On Aldabra Island, off the east coast of Africa, all the numerous flightless birds except one have disappeared since the cat came, and that one exists now in numbers only on some smaller islands of the group that the cat has not reached.[3] On Glorioso Island numbers of cats range the jungle, and birds have been decimated even more than on Aldabra.

A few cats often are enough to destroy the birds on a small island. The cats get the birds in the nesting season when incubating eggs or brooding young, and thus prevent breeding. A cat belonging to Peter Lyall, the lighthouse keeper on Stevens Island (a wooded island hardly a square mile in extent in Cook's Strait), exterminated a little wren, *Traversa lyalli*. Only twelve specimens are now in existence, and all these were brought in by this cat, an excellent hunter, which roamed over the entire island. How many more she ate or left dead in the woods will never be

---

[1] Chapman, Frank M.: Camps and Cruises of an Ornithologist, 1908, p. 39.
[2] See Rothschild, Walter: Extinct Birds, 1907, pp. 21, 128; also Forbes, Ibis, 6th series, V, 1893, p. 523.
[3] Abbott, W. L.: Proceedings of the National Museum, Vol. XVI, 1893, pp. 762, 764.

known. It is believed that this bird lived formerly on d'Urville Island and even on New Zealand itself, where cats had been introduced many years before.[1] Dr. Louis B. Bishop of New Haven writes me that in 1901-02 he found the piping plover and Wilson's plover breeding "tolerably commonly" and Virginia rails and Clapper rails abundantly on Pea Island, N. C., but in December, 1908, Mr. J. B. Etheridge, manager of the club on the island, told Dr. Bishop that the piping plover had been exterminated, Wilson's plover almost extirpated and rails greatly reduced by cats from the Pea Island life-saving station. The station was closed in summer and the cats were abandoned.

Mr. Wilbur F. Smith of Norwalk, Conn., visited Wooden Ball Island, off the coast of Maine, where there was a colony of Leach's petrels. He found that the entire colony had been destroyed. Passing by one of the fishermen's cabins he noticed the ground strewn with petrels' remains, some freshly killed. The fisherman told him that the cats caught the birds at night and brought them to the house to eat; he said that there were but three cats kept and only one wild house cat had been seen. A great colony of petrels on Great Duck Island has been decimated in recent years by a few cats kept there by the lighthouse keepers.

Several years ago the least tern was very nearly exterminated in New England by milliners' agents, but finally, by a stringent enforcement of the law, they were saved from extinction. In 1907 a considerable number established themselves not far from the lighthouse on Monomoy, at the elbow of Cape Cod, but the birds could not rear young on account of cats which roamed the beach. I visited the place in 1908 and found that the colony had been broken up, and that the beach was pitted with many cat tracks.

Space will not allow many details of the cats' destructiveness to birds on islands, but there is room for the sequel to the story told by Mr. G. K. Noble in the "Warbler," of Sept. 1, 1913. He asserted that on the south end of Muskeget Island a great Massachusetts colony of sea birds protected by the town of Nantucket, the breeding gulls and terns, had been nearly extirpated by cats. Mr. Howard H. Cleaves wrote me in 1914 that the warden in charge said that if the cats continued to increase they would exterminate the entire colony of some 45,000 birds within five years. All over that part of the island that the cats mostly inhabited could be seen the uneaten bodies of terns killed on their nests, their heads torn off, and the wings and feathers

---

[1] Rothschild, Walter: Extinct Birds, 1907, p. 25.

of those that had been eaten. The mangled bodies of newly hatched young, as well as larger young, were found scattered about profusely. There are no trees on the island, therefore hawks and owls do not nest there, and do not remain there during the nesting season of the birds. There are no predatory mammals except the cat, and the indigenous short-eared owl was exterminated years ago. Therefore the cat is practically the only enemy with which the gulls and terns have to contend. Mr. Arthur Brigham of Boston wrote me in 1914 that the cats had greatly depleted the number of the birds, and an agent of the Nantucket Society for the Prevention of Cruelty to Animals reported the same year that in a brief search he found fully a thousand nest sites with the remains of parent birds, egg shells and young scattered about them. Whether the cats increased or not we do not know, but during the summer of 1914 it was easy to gather a bushel of wings of the dead birds. The warden killed three cats in 1913, and may have destroyed a few in 1914, but Deputy Fish and Game Commissioner William Day went to the island in the winter, and, with a good dog, found and shot seven cats, one of them a female heavy with young; another cat was found dead. Mr. Day believes that he has killed every cat there, and the dog could find no more. This shows clearly how terribly destructive a few stray cats can be among breeding birds, and how they kill, not merely to eat, but for the love of killing. Since the above was written Mr. W. L. McAtee of the Biological Survey has informed me that more cats have been let loose on the island by fishermen, and that the number of birds was much reduced by them in 1915.

Expert Opinions on the Cat's Destructiveness to Birds.

In all my investigations into the economic status of the cat, opinions have been disregarded and only facts sought. Nevertheless, opinions of all kinds have been offered. Many cat lovers naturally are loath to believe or admit that their pets seriously menace the birds, but some frankly avow the regrettable facts. Miss Helen Leighton, president of the Animal Rescue League of Fall River, writes: "I have found the cat a beautiful, clean, intelligent and affectionate pet, readily trained not to molest cage birds, but also a very dangerous enemy to bird life in general. It is idle to deny the latter point." Miss Mary A. White of Heath writes: "I am fond of cats and consider them a close and valuable bond, endearing animals to humans, but do not keep one because I have found them so destructive to bird life."

Dr. C. F. Hodge, author of "Nature Study and Life," and an authority on the rearing of game birds, says that evidence from all civilized countries in which measures are being taken to protect game and insectivorous birds is overwhelming that the cat is the worst enemy of bird life. Most authorities lean toward this opinion.

If opinions are to be regarded at all, those of well-known, conservative people who have made a lifelong study of birds, their enemies and the means of protecting them should be entitled to the greatest weight, as such people, interested in the protection of birds, are best qualified to express an opinion by reason of long experience and habits of close observation.

Mr. Witmer Stone of the Philadelphia Academy of Sciences, editor of the "Auk," and for many years chairman of the American Ornithologists' Union Committee on Bird Protection, writes: "There is, I think, no doubt that for years past the greatest destructive agency to our smaller song and insectivorous birds has been the cat."

Robert Ridgway, of the Smithsonian Institution at Washington, D. C., whose monumental standard works on American ornithology are known throughout the world, writing of roaming cats in the locality of his home in southern Illinois, says: "It is of course difficult to estimate the extent to which these practically wild cats are responsible for the present relative scarcity of birds, but it must, from the very nature of the case, be a most important factor."

John Burroughs says that cats probably destroy more birds than all other animals combined. He believes that the preservation of birds involves the nonpreservation of cats.

Dr. Frank M. Chapman, of the American Museum of Natural History, author of standard works on American ornithology and editor of "Bird-Lore," has this to say on the subject: "The most important problem confronting bird protectors to-day is the devising of a proper means for the disposition of the surplus cat population of this country. By surplus population we mean that very large proportion of cats which do not receive the care due a domesticated or pet animal, and which are, therefore, practically dependent on their own efforts for food."

Mr. Henry W. Henshaw, chief of the Biological Survey, United States Department of Agriculture, says that one of the worst foes of our native birds is the house cat. Probably none of our native wild animals destroy as many birds on the farm, particularly the fledglings, as do cats.

Mr. William Dutcher, president of the National Association

— of Audubon Societies, considers the wild house cat one of the greatest causes of bird destruction known. He says that the boy with the air gun is not in the same class with the cat.

Dr. William T. Hornaday, director of the New York Zoölogical Park, and author of valuable works on the protection of wild life, says: "In such thickly settled communities as our northern States, from the Atlantic coast to the sandhills of Kansas and Nebraska, the domestic cat is probably the greatest four-footed scourge of bird life. Thousands of persons who never have seen a hunting cat in action will doubt this statement, but proof of its truthfulness is only too painfully abundant. . . . That cats destroy annually in the United States several millions of very valuable birds seems fairly beyond question. I believe that in settled regions they are worse than weasels, foxes, skunks and mink combined, because there are about one hundred times as many of them, and those that hunt are not afraid to hunt in the daytime. Of course, I am not saying that all cats hunt wild game; but in the country I believe that fully one-half of them do."

Mr. T. Gilbert Pearson, secretary of the National Association of Audubon Societies, and author of books and papers on birds, makes the following statement: "There is no wild bird or animal in the United States whose destructive inroads on our bird population is in any sense comparable to the widespread devastation created by the domestic cat."

Dr. George W. Field, chairman of the Massachusetts Commission on Fisheries and Game, while fond of cats as pets, says that he has reluctantly concluded that they destroy more game and insectivorous birds than any other one factor at present operating to diminish the bird population.

Mr. Ernest Harold Baynes, author of "Wild Bird Guests," etc., regards the cat as "far and away" the most destructive of all the animals for whose present status as bird destroyers man is more or less responsible.

Mrs. Mabel Osgood Wright, president of the Connecticut Audubon Society, and author of many popular books on birds, writes: "The evidence of men and women whose words are incontestable would verify my most radical statement, but one fact is beyond dispute: if the people of the country insist upon keeping cats in the same numbers as at present, all the splendid work of Federal and State legislation, all the labors of game and song bird protective associations, all the loving care of individuals in watching and feeding, will not be able to save our native birds in many localities."

Mr. Henry Nehrling, a well-known writer on American birds and bird protection, goes so far as to say: "They do more harm to our familiar garden birds than all other enemies combined." Baron Hans von Berlepsch, perhaps the greatest authority on bird protection, asserts: "We may as well give up protection of birds about our gardens and houses, so long as we tolerate cats outside the buildings;" and concludes: "Therefore, against all cats found loitering outside of buildings, the most relentless war of extermination."

Destruction of Mammals and Lower Animals by Cats.

During such research as I have been able to make through the literature of the subject, it has become evident that naturalists and writers on rats and ratcatching, and writers on sport and gamekeeping, almost invariably belittle the cat as a ratcatcher, but admit that it catches many mice and much game. Even the health authorities in various countries who have had to take up rat destruction in the seaports of the world in order to check the bubonic plague do not, as a rule, seem to appreciate the cat's assistance. Occasionally one is found who gives the cat credit for good work, but this is the exception, and I find very little evidence anywhere that cats destroy other predatory animals. No one of my correspondents records a hawk, fox, raccoon or mink as killed by a cat. One records *one* attack on a skunk. It was not repeated. Three tell of weasels killed by cats, one of a woodchuck and one of a muskrat, but the harmless or useful mammals appear to be killed in great numbers; also squirrels and rabbits.

Squirrels.

We find that 196 observers report many squirrels killed by cats. Mr. William Brewster says that almost all the chipmunks, most of the red squirrels and many gray squirrels are killed annually wherever cats roam freely and numerously. Cats have exterminated the chipmunks on my farm, but have not been numerous enough to make much impression on the numbers of the more arboreal squirrels. I have seen cats carrying very large gray squirrels, but the larger ones will sometimes whip a cat and drive it away.

Hares and Rabbits.

The number of observers reporting that cats kill many rabbits is 149. The majority of these rabbits (hares) are young cottontails, but many adults are killed, and some of the larger northern varying hares or "white rabbits," so called. The cat is so destructive to rabbits that on Sable Island, off the coast of Nova Scotia, which had been the home of these little animals for at least half a century, the introduction of a few cats was followed by the absolute extinction of the rabbit population. There is abundant evidence of the rabbit-killing habit. Mr. Cassius R. Tirrell of South Weymouth tells of a cat that brought home 7 young rabbits in two days. Mr. Albert E. Shedd of Sharon writes that he had a cat in 1910 that killed many rabbits, grouse and some small birds; it brought in 4 cottontails in a single day. Mr. A. K. Learned of Gardner tells of a cat that brought in 22 rabbits in one summer. Jones and Woodward record the confession of a lady in a local paper that her cat, with kittens, brought in in one week 26 mice, 19 rabbits, 10 moles, 7 young birds and 2 squirrels, and they say that they have heard of cats "a great deal worse."[1]

Dr. William T. Hornaday tells in his interesting and useful book, "Our Vanishing Wild Life," how in one year cats killed nearly all the wild rabbits in the park — some eighty or ninety. The cats were exterminated, and the rabbits slowly increased. Several observers have reported a cat going out at dusk and returning in a few minutes with a full-grown rabbit. My friend, William C. Peterson of Canaveral, Fla., saw his cat kill one. This cat frequently brought in adult cottontails, and its owner desired to see how it overcame them. One evening, when he saw one sitting in his garden, he took the cat out there. She sprang on the rabbit, caught it with her teeth by the back of the neck, and lying beside it caught its haunches with her hind claws and straining hard stretched and apparently broke its neck. It was all over in a moment.

Moles and Shrews.

Cats kill many moles. This is reported by 132 observers. Only 51 say that cats kill many shrews. Evidently many observers do not distinguish shrews from moles. Others admit that they do not know the shrew. The short-tailed shrew, *Blarina brevicauda*, closely resembles a mole in appearance, while some of the smaller shrews might be mistaken for mice by the casual observer. Cats kill considerable numbers of moles and shrews, but they rarely eat them, as there seems to be some disagreeable scent or taste about them.

---

[1] Jones, Owen, and Woodward, Marcus: The Gamekeeper's Notebook, London, 1910, pp. 263, 264.

Rats and Mice.

Many statements have been published recently to the effect that not one cat in fifty or even one in a hundred kills rats. These statements are at variance with my experience, as well as with that of most of my correspondents, and they cannot be founded on any careful investigation. Nevertheless, it is true that many cats do not hunt rats. Dr. A. K. Fisher, in charge of the economic investigations of the Bureau of Biological Survey, United States Department of Agriculture, says: —

It is impossible at present to obtain correct figures on this subject, but it is safe to say that few persons in a normal lifetime run across more than half a dozen cats that habitually attack rats. Occasionally a hunter-cat is found which seems to delight in catching rats, gophers or ground squirrels. It has been the common experience of the writer to find premises that were well supplied with cats overrun with rats and mice. At a certain ranch house in the west, he trapped twelve mice in his bedroom in a week, although eight cats had access to the place.[1]

Dr. G. M. Corput, another Government expert, in the United States Public Health and Marine Hospital Service, gives an experience which seems to show that little dependence can be placed on the cat as a rat exterminator.

Every quarantine officer is familiar with the old plea of shipmasters that there is no use of fumigating the cabin of a vessel because there is a cat on board which is an excellent ratter and renders it impossible for rats to live in cabin. The enclosed pictures are the result of not believing this story. The British steamship "Ethelhilda" arrived at this station [New Orleans Quarantine] March 18, from the west coast of Africa. The captain assured me that it was impossible for any rats to be in the cabin of his vessel because of the presence of an exceptionally good cat. The cabin was nevertheless fumigated. Through the irony of fate the cat was forgotten. Then the cabin was opened, and the enclosed picture shows the result. Every part of the ship had many rats. The picture is limited, however, to what was found in the cabin, one cat, twenty-four rats.[2]

In my experience of forty years only two of my own cats have habitually attacked rats. Most of them did not trouble rats at all, a few got one occasionally, but the best one on the farm killed on the average about one a week, or over fifty a year. Upon the arrival of this cat, the rats soon disappeared and were not seen running about as before. A little careful investigation, however, showed that they were nearly as numerous as ever, but much shyer, keeping out of sight. At the end of the year,

---

[1] Fisher, A. K.: Yearbook, U. S. Dept. of Agr., 1908, pp. 189, 190.
[2] Public Health Reports, Vol. 29, No. 16, April 17, 1914, p. 928.

notwithstanding the killing done by the cat, the number present had not decreased, as not enough had been killed to dispose of the annual increase. After the cat had been in the barn six months, I set eleven old rusty traps one night and got six rats; two sprung traps and got away. This one night's work of old and rather ineffective traps equalled six weeks' work of the cat. No one knows how many rats infest his place when he keeps a ratcatching cat, for then the rats almost invariably keep out of sight. I have found it difficult to get rid of rats when I had cats, as traps could not be set freely on account of the cats, but as soon as the cats were disposed of, the rats were trapped. I have just returned from a visit in the country with a friend who keeps two cats which, he says regretfully, are very destructive to birds. When asked why he did not dispose of them he replied that a farmer must have cats to catch the rats and mice about his buildings. At that very moment there were two traps set for mice in a living room, and he admitted that whenever rats became unbearable in his barn the cats were shut out and poison was used. Apparently, however, my own experience with cats has been unfortunate, as the farm canvass undertaken by the State Board of Agriculture shows that about four-fifths of the farmers interviewed seem to believe that cats are more or less effective as rat killers. The following figures are given for what they are worth. They refer to village and farm cats: —

Interviews,	291
Cats kept,	559
Known ratters,	197
Known not ratters,	43
Have rats,	118
Have no rats,	131
Had more rats before getting cat,	22
Have rats and no cat,	27
Have no rats and no cat,	24
Have cats and no rats,	107
Have both cats and rats,	96

89 keeping 184 cats use traps also.
45 keeping 90 cats use poisons.
36 keeping 70 cats use both traps and poisons.

These figures, furnished mainly by friends and owners of cats, do not speak highly of the ratcatching ability of the average cat, but they seem to show that more than one-third of the cats kept by these country people kill more or less rats. A little more than one-fifth seem to be effective ratcatchers, as they appear to have killed or driven out rats. It is safe to say that some of the people who asserted they had no rats really had them at the time, although they did not realize it, as there are many more rats than are seen by human eyes. Mr. McMahon, in this canvass, found a village where quantities of fowls were kept and where cats were depended on to exterminate the rats. Everybody there seemed to believe that cats were effective as rat exterminators, and no one seemed to be using traps or poisons. The village was canvassed quite thoroughly, and every place was found infested by rats, while in nearly every place cats were kept. The evidence did not confirm the popular belief in the cat.

These statistics were taken in summer and early fall, before the rats began to come into buildings for the winter. A census taken in December probably would have revealed a larger number of places infested. Most people are not anxious to admit that there are rats in their dwellings. The above facts considered, it is probable that some of the figures given unduly favor the cat.

Turning now to the observers who filled out the questionnaire, a large part of whom are town or city people, we find the following: —

Reports,	324
Keep cats,	99
Number kept,	132
Do not keep cats,	225
Average number of cats per family of correspondents keeping cats,	1.3
Cats per family in neighborhood, reports,	360
Total number of cats on these reports,	515
Average number of cats per family in neighborhood,	1.4
Rats numerous,	78
Rats common,	151
Rats rare,	137
Rats have decreased since cats were obtained,	164
Rats have not decreased since cats were obtained,	94
Believe cats exterminate or drive out rats,	71
Believe cats do not exterminate or drive out rats,	221
Mice have decreased since cats were obtained,	190
Mice have not decreased since cats were obtained,	71
Believe cats exterminate or drive out mice,	84
Believe cats do not exterminate or drive out mice,	217
Have both cats and rats,	65
Cats kept as pets alone,	84
Cats kept as mousers,	39
Cats kept as both pets and mousers,	169

A typographical error in the questionnaire makes it impossible in most cases to get the maximum number of rats or mice killed by a cat in one day, as the question regarding rats reads, "How many rats have you known to be killed by cats in a day?" Hence a reply may include two or a dozen cats. In a few cases, however, it is stated specifically that one cat killed a certain number. Only 147 out of 427 observers can say that they ever knew cats to kill any definite number of rats in a day. In most cases the maximum number of rats killed by cats in a day varies from 1 to 3, but Mr. B. S. Bowdish of Demarest, N. J., records 5 small rats killed. There are a few cases where larger numbers are given. Miss Grace E. Wilder of East Lynn has a cat that has killed 4 rats in a day. Mr. Jonathan H. Jones of Waquoit records 7 to one cat. Mrs. Mary A. Wheat of Dorchester has known a cat to kill 14. Mr. F. H. Mosher of Melrose has a cat which killed 18 in one day, 15 of which were young. When grain is being cleared out of a building, a good ratter occasionally makes a great killing. Mrs. Florence G. Butler of East Charlemont says that she has known cats thus to kill 20 rats in a corn barn. An enthusiastic friend of the cat wrote that she had known 32 rats killed by a cat in one day, and that another averaged 10 rats a night, which would amount to 3,650 rats per year; she also speaks of another cat which was alleged to have killed enough field mice nightly to "cover" the doorstep and the walk leading up to it. Such destruction as alleged here would soon solve the rat problem. The first of her stories was investigated, with the following results: —

A porter of a large dry goods house gave a signed statement, saying that the first cat mentioned, which he had obtained from the Animal Rescue League of Boston, killed 32 rats between Saturday night and Monday night, and that another averaged from 3 to 5 a night. An investigation of this statement showed that in the first case heads, tails and other remains of rats were counted, and that there were two cats instead of one. The man who now cares for this champion cat has never known it to kill more than 7 rats in one night.

Miss Clara L. Hutchins of Groton has three cats that are regarded as excellent rat killers. At my request she kept a careful record, with dates, of the rats killed by them from June 28 to September 1. Teddy killed 4, 2 of which were full grown. Buster killed 6, 2 of which were full grown. Binks killed 9 small rats. By actual count, here were 15 small rats and 6 full-grown ones killed by these excellent cats in a little over two months; and it is quite possible that a few more may have been killed, as the remains of 2 more were found. The record also gives 2 mice and 3 small snakes, all killed by Buster. It is probable that few actual records carefully kept would show better results than this, except possibly where rats swarm. Mr. Wilfrid Wheeler, secretary of the Massachusetts State Board of Agriculture, had a cat which, he says, caught about 2 rats a day for two weeks, but the rodents were so plentiful that this cat's work made no apparent difference in their numbers and destructiveness, and it was found necessary to resort to poison. Dr. George W. Field of Sharon has found traps, poisons, terriers and other means necessary with rats, even on a farm where ten to twelve cats were kept.

The evidence of my hundreds of correspondents regarding the value of the cat as a ratcatcher is varying and contradictory. Many correspondents find their cats very useful in reducing the numbers of rats in barns and outhouses, or in driving them from dwellings and poultry houses. Many others find theirs absolutely worthless for these purposes. On a farm where there were several cats, the farmer was anxious to know about the best rat traps, as the premises were overrun with rats, and they had entered the bird cage and eaten the canary. A poultryman said that rats swarmed all over the place, although there were so many cats there that he could not give the exact number. A miller asserted that cats were short-lived in his mill as the rats were too much for them. Another had a cat that kept his mill nearly free from rats and mice. There are many tales of cats beaten, cornered and even killed by full-grown rats, and others of cats that are believed to have killed large numbers of rats with impunity, all of which goes to show that there is much difference in cats.

In speaking of mice there is more agreement; although some cats will not touch mice, the majority apparently catch them. This has been the experience of mankind for centuries, but as mice are easily caught by any one with energy enough to set mouse traps, the principal advantage of the cat as a mouse trap is that it is "easy to set." Any intelligent, observing, persistent person can exterminate mice with traps, except perhaps in granaries and like buildings where abundant food is accessible. Cats, on the other hand, cannot exterminate them as they sometimes extirpate rabbits, for the reason that they cannot follow mice into their holes, and they cannot, like traps, attract them from their holes. Nevertheless, a good mouser often will make life so unpleasant for mice, as well as rats, that they will leave a dwelling house inhabited by such a cat and go where cats are not kept. Such cats are valuable, if they can be confined to the premises where rats and mice are troublesome.

Bats.

Probably not very many bats are caught by cats as compared with the number of birds destroyed, for bats never willingly come to the ground. Occasionally a low-flying bat is struck down by a cat, or one that has entered a dwelling house is caught, but only two observers report to me the destruction of bats by the cat.

Reptiles and Amphibians.

As the hunting cat strikes practically every quick-moving object it can reach and master, toads, frogs, lizards, newts, salamanders and snakes, particularly the useful, smaller species, are decimated. Many cats destroy the beneficial toad at night, when it is most active, while frogs are less often molested. The killing of toads by cats is done mainly under the cloak of darkness, but I have seen cats killing them under the street lights at night. Four observers report cats killing toads, and five have observed them killing frogs.

Fish.

The well-known antipathy of cats for water would seem to preclude fishing as a feline accomplishment, but five of my correspondents report fishing cats. In two cases the identity of the fish caught could not be determined. In other cases, trout, smelt and eels were caught. Mr. E. Colfax Johnson says that when the streams are low in summer, cats get many trout. This is corroborated by others. Mr. James E. Bemis of Framingham has seen cats catching smelt in shallow pools left by the receding tide. One cat "flipped" out three with her paw and carried all three away in her mouth at once. Cats may get the fishing habit at the seashore or by finding fish dead or dying.

The fishing cat.

Stables says that a cat may be easily taught to fish by taking her, when young, to a shallow stream on a clear day when minnows are plentiful, and throwing in a few dead ones, meanwhile encouraging her to catch them, when she will soon learn to catch the living fish.[1] Buckland, Darwin and others tell of cats which, without teaching, learned to go into the water and catch fish. Stables asserts that he has "dozens" of well-authenticated anecdotes of cats expert at fishing. He avers that he watched one dive into a stream and emerge almost immediately with a large trout in its

---

[1] Stables, Gordon: The Domestic Cat, 1876, p. 91.

mouth. He says that cats spring off the bank and dive, not only in catching fish, but in pursuit of water rats, and that in Scotland cats often attack salmon and destroy large quantities in small streams in the spawning season. Millers' cats, and cats living near streams, by the sea or by artificial fish ponds are the chief offenders.[1]

Crustaceans and Mollusks.

Dr. A. K. Fisher asserts that he once saw a cat in a fisherman's house on the south shore of Long Island, N. Y., that caught crabs by wading out into the water for them. Both salt-water and fresh-water clams and even oysters are eaten by cats.

Insects.

Cats strike down and kill some large insects and a few of the smaller species, particularly those of the fields, such as moths, May beetles, grasshoppers and crickets. Occasionally a cat makes a business of catching and eating grasshoppers, but apparently the animal is not naturally insectivorous, as many observers agree that puss grows thin on such a diet. Prof. H. A. Surface asserts that he observed a cat pouncing on crickets and grasshoppers in the grass, and that one ate so many May beetles or "June bugs" that it threw up "nearly a pint"

The insect killer.

of the "outer shells" of these beetles. Many report that cats sicken on an insect diet, but they probably disgorge the hard and indigestible parts of insects, as do many birds. Probably the insect food of cats ordinarily is an unimportant part of their regimen, but insects may serve to fill the stomach when sufficient animal food of other kinds is lacking. Following is a compilation from many reports: —

Species killed.	Number reporting it.	Species killed.	Number reporting it.
Grasshoppers,	169	Locusts,	11
Crickets,	69	Ants,	5
Flies,	41	Water bugs,	4
Moths,	29	Bees,	2
Beetles,	24	Wasps,	1
Butterflies,	26	Hornets,	1
"June bugs,"	15	Katydids,	1

[1] Stables, Gordon: The Domestic Cat, 1876, pp. 161, 162.

THE ECONOMIC VALUE OF THE CAT.

Economic Value of Weasels.

— The destruction of weasels must count against the cat in so far as it removes from the field the most effective mammal enemy of rats and mice. Weasels are ravenous, persistent slayers of small rodents, and are able to follow them into all their holes and hiding places; but unfortunately the food habits of the weasel in this country are not well enough known to enable one to speak with authority regarding its depredations on insect-eating birds and other insectivorous creatures. Occasionally it kills fowls and game birds, and it is regarded as vermin by the farmer and gamekeeper. Probably cats do not kill many weasels and their destruction need not be given much weight.

Economic Value of Squirrels.

— The killing of squirrels by cats will be regarded by farmers generally as a beneficial habit, as squirrels are destructive to fruit and grain. Sometimes they destroy eggs and young birds; but the cat kills mainly chipmunks, which are least destructive to fruit, grain and birds, although many red squirrels and a few grays are taken. Cats undoubtedly save the lives of some birds by killing squirrels, but, on the other hand, they thus protect many insects, probably as many as cats themselves destroy. I watched a gray squirrel with a glass and saw it go thoroughly over an oak tree about forty feet high, gleaning nearly all the insects upon it. Mr. C. A. Lyford reports that he watched a red squirrel take all the bark lice from a large section of the trunk of a white pine. Mr. W. L. Burnett, Prof. C. P. Gillette and Prof. J. M. Aldrich, reporting on examinations of the striped ground squirrels or spermophiles, find that they eat quantities of injurious insects, such as caterpillars, including cutworms and webworms, grasshoppers, locusts and ground beetles. Grasshoppers seem to be preferred to all other food. Cutworms are eaten in numbers.[1] Mr. Walt F. McMahon informs me that squirrels gnaw into the burrows of the leopard moth and extract the larvæ. Most insects eaten by squirrels are injurious and squirrels kill and eat some mice.

The food of New England chipmunks is believed to include many injurious insects. The destruction of these little animals by the cat may be at times an injury and at other times a benefit

---

[1] Burnett, W. L.: Circulars Nos. 9 and 14, issued from the office of the State Entomologist, Fort Collins, Col.

to the farmer. The value of the gray squirrel as a game animal is considerable. Therefore, whether the destruction of squirrels by cats is beneficial or injurious to mankind will depend largely upon the circumstances and the point of view, and need not be given great weight; but the cat may be serviceable toward checking the undue increase of squirrels where their native natural enemies are not numerous, for in such cases squirrels become very destructive.

Economic Value of Hares or Rabbits.

The destruction of hares (or rabbits, so called) by cats may be placed in the same category. Hares often become injurious by gnawing the bark of fruit trees, and as they are vegetable feeders they are not looked upon with favor by the farmer. But from the standpoint of the sportsman they form collectively a valuable asset to any land, and their food value is too great in these days, when meat is high in price, to make them economical as food for cats.

Economic Value of Moles.

Moles often become nuisances in mowing lands and on lawns, where they throw up unsightly ridges and mounds; also in gardens they disturb the roots of plants by their digging; but careful investigation shows that they are very rarely vegetable feeders, and that the destruction of plants sometimes attributed to them by farmers is caused not by moles but by mice, which sometimes use their burrows. Every subterranean mole gallery forms a trap into which worms and grubs continually tumble, and the mole, moving rapidly through its tunnel at all hours of the day and night, gathers them in. It is one of the chief enemies of the white grub of the May beetle; also of wireworms, the progeny of snap beetles, both of which are destructive to the roots of grass and cultivated plants, and are difficult to control. The reason that mole burrows often follow rows of vegetables is that the mole is seeking grubs at the plant roots. The moles killed by cats, had they been allowed to live, would have eaten an enormous number of injurious insects, — far more than cats would ever kill.

Economic Value of Shrews and Bats.

The killing of many shrews by cats forms one of the blackest pages of the record, for there are few creatures so harmless and so beneficial as the shrew, from the standpoint of the agriculturist. Shrews are tremendous gluttons and feed very largely

— on insect life. Apparently they never touch the products of man's labor. The species most commonly killed by cats in Massachusetts is the short-tailed shrew, *Blarina brevicauda*. This little mammal probably is mistaken for a small mole by most people, as it somewhat resembles the common mole.

Mr. John Norden believes that this gluttonous animal eats about twice or three times its own weight in twenty-four hours,[1] but probably this is exceptional. Nevertheless, the shrew requires an amount of food equal to nearly its own weight daily, and cannot live long without food. It destroys enormous quantities of worms and insects, and kills many field mice and other mice larger than itself. Shrews may kill more field mice annually than cats destroy. Mr. H. L. Babcock, who has studied the shrew, considers it of great economic value.[2] In killing these shrews, therefore, the cat protects quantities of insects and mice which these shrews and their numerous progeny might otherwise destroy.

New England bats are remarkably useful creatures, as they subsist on mosquitoes and other nocturnal insects which often escape the birds by day, and thus they fill a gap which can perhaps be filled by no other creature. Apparently they have no harmful habits, and their destruction must be set down as against the cat.

Economic Value of Amphibians and Reptiles.

The smaller snakes and the toads, frogs, salamanders, newts and lizards which are destroyed by cats all have been proved to be practically harmless and very beneficial as destroyers of insects.

The toad is an example of the beneficial character of the amphibians. Kirkland finds that the food of the common toad is practically all of an animal nature. Ants form 19 per cent; cutworms, 16 per cent; tent caterpillars and other injurious leaf-eating caterpillars, 12 per cent; June beetles, potato beetles, snap beetles, weevils and allied beetles make up 18 per cent; snails, thousand-legged worms, sowbugs and other injurious forms compose 14 per cent; supposedly beneficial species, such as ground beetles, spiders and carrion beetles, make up 11 per cent, and there is 2 per cent of vegetable and mineral matter, probably taken incidentally with the animal aliment. The food of the toad, therefore, appears to consist mainly of 81 per cent of injurious species, against 11 per cent beneficial ones. The remainder is unidentified animal [insect?] food.

---

[1] Canadian Sportsman and Naturalist, Vol. III, 1883.
[2] Babcock, H. L.: The Food Habits of the Short-tailed Shrew, Science, new series, Vol. XL, No. 1032, pp. 526-530.

The capacity of the toad is enormous. A single stomach contained 77 myriapods or thousand-legged worms; another, 37 tent caterpillars; a third, 65 caterpillars of the gypsy moth; and a fourth, 55 army worms. Individual toads have been seen to eat as follows: No. 1, 30 full-grown celery caterpillars; No. 2, 86 house flies; No. 3, 90 rose bugs.[1]

The toad is a highly beneficial animal and should be protected by law and public sentiment. Every toad killed by a cat is much more useful as an insect destroyer than the cat which kills it. When we consider that practically all our frogs, lizards, salamanders and little snakes are insectivorous and harmless, and differ from the toad mainly in the degree of their utility and in the fact that some feed by day rather than by night, we can see that the cat which kills such harmless, useful creatures is likely to work much injury to the agriculturist.

For an investigation of the food of the amphibians, see the first report on the economic features of the amphibians of Pennsylvania, by H. A. Surface (Bi-monthly Zoölogical Bulletin of the Division of Zoölogy of the Pennsylvania Department of Agriculture, Vol. III, Nos. 3 and 4, May–July, 1913).

Economic Value of Birds.

The killing of birds is the most serious item in the account against cats, except possibly their agency in the dissemination of disease. All birds smaller than geese, including domestic fowls and excepting birds of prey, are in danger of being attacked and killed by cats, which habitually kill birds up to the size of a pigeon. The birds destroyed by farm cats and house cats are mainly of the species that are most common and useful about gardens, orchards and fields, while vagabond cats and woods cats destroy the most valuable of the woodland birds and game birds. The list includes all that nest and live upon or near the ground, all that feed there, and most of those that nest and feed in trees, as they have to come to the ground to drink and bathe. The following list of 107 species of birds killed by cats is compiled from the papers of correspondents, and while it does not include all the species attacked in Massachusetts, it includes most of the genera: —

---

[1] Kirkland, A. H.: The Garden Toad, Massachusetts State Board of Agriculture, Nature Leaflet No. 28, fourth edition, December, 1913.

Species of Wild Birds Reported Killed by Cats.

Name of Bird.	Number reporting it.	Name of Bird.	Number reporting it.
Bluebird,	75	"Vireos,"	11
Robin,	272	Yellow-throated vireo,	1
"Thrush,"	16	Warbling vireo,	1
Hermit thrush,	5	Red-eyed vireo,	3
Olive-backed thrush,	1	Shrike,	1
Veery,	2	Cedar waxwing,	9
Wood thrush,	2	Tree swallow,	7
Ruby-crowned kinglet,	4	Barn swallow,	42
Golden-crowned kinglet,	1	Cliff swallow,	3
Chickadee,	24	Purple martin,	4
White-breasted nuthatch,	6	Scarlet tanager,	2
Brown creeper,	3	Indigo bunting,	2
Long-billed marsh wren,	1	Rose-breasted grosbeak,	7
Winter wren,	3	Cardinal,	1
House wren,	16	Towhee,	6
Carolina wren,	1	Fox sparrow,	3
Brown thrasher,	14	Lincoln's sparrow,	1
Catbird,	52	Song sparrow,	46
Mockingbird,	2	Slate-colored junco,	34
"Warblers,"	17	Field sparrow,	10
Redstart,	3	Chipping sparrow,	54
Canadian warbler,	1	"Sparrows,"	29
Wilson's warbler,	2	Tree sparrow,	5
Yellow-breasted chat,	2	White-throated sparrow,	5
Northern yellow-throat,	1	Seaside sparrow,	1
Kentucky warbler,	1	Henslow's sparrow,	1
Water thrush,	1	Grasshopper sparrow,	1
Oven-bird,	6	Savannah sparrow,	3
Black-throated green warbler,	1	Vesper sparrow,	8
Blackburnian warbler,	1	Snow bunting,	1
Blackpoll warbler,	2	Goldfinch,	14
Chestnut-sided warbler,	2	White-winged crossbill,	1
Magnolia warbler,	1	Purple finch,	2
Myrtle warbler,	9	English sparrow,	72
Yellow warbler,	20	Pine grosbeak,	3
Nashville warbler,	1	Grackle,	11
Black and white warbler,	3	Blackbird,	5

SPECIES OF WILD BIRDS REPORTED KILLED BY CATS — *Concluded.*

Name of Bird.	Number reporting it.	Name of Bird.	Number reporting it.
Baltimore oriole,	13	Screech owl,	2
Meadowlark,	15	Saw-whet owl,	1
Red-winged blackbird,	5	Mourning dove,	2
Bobolink,	7	Heath hen,	1
Starling,	2	Ruffed grouse,	46
Crow,	1	Ring-necked pheasant,	11
Blue jay,	25	Golden pheasant,	1
"Flycatchers,"	3	Hungarian partridge,	1
Least flycatcher,	2	Bobwhite,	44
Wood pewee,	2	Spotted sandpiper,	1
Phœbe,	9	Woodcock,	11
Kingbird,	5	Yellow-legs,	1
Ruby-throated hummingbird,	10	Gallinule,	1
Chimney swift,	8	Yellow rail,	1
Nighthawk,	3	Sora,	2
Whip-poor-will,	3	Virginia rail,	4
Northern flicker,	24	"Rail,"	1
"Woodpeckers,"	8	Black-crowned night heron,	1
Downy woodpecker,	7	Leach's petrel,	1
Cuckoo,	2	Dovekie,	1

This list would seem to indicate that more robins than any other species are killed by cats. In the cities, where the so-called "English" sparrow is more plentiful, it suffers considerably, though not so much as the robin, for it can take better care of itself, having lived with the cat for many centuries. Therefore, against 272 observers reporting the robin, we have only 72 noting the English sparrow, but there are 29 reporting sparrows without noting the species, some of which probably were "English." If the ravages of the cat were confined to the robin and the introduced sparrow they might be borne, as the sparrow, like the cat, is a foreign disturber, and the robin, like the sparrow, is so fecund that when protected it makes good its losses. But when such useful birds as the native bluebirds, chickadees, cuckoos, sparrows, swallows, thrushes, titmice, wrens, warblers, woodpeckers and meadowlarks are included in the great toll that the cat takes from bird life, the matter becomes really serious.

It is a well-known fact that since the settlement of the United States, insect pests and the injury done by them have increased constantly. It is well known also that birds destroy enormous numbers of insects, and that many species of birds have been reduced greatly in numbers, while some have been exterminated. Both the destruction of birds and the increase of insect pests have been greatest within the last century. This is more than a mere coincidence. Many smaller useful species probably increased when the forests were cleared from the Atlantic coastal plain, farms established and fruit trees planted, but their increase has not kept pace with the multiplication of insect pests, on which they feed, and the domestic cat has been one of the chief factors in keeping down their numbers.[1] As the population increases, cats increase. Birds are not nearly so plentiful in Massachusetts to-day as they are in some western States, and their numbers compare very unfavorably with those in older countries, like England and Germany, where stray cats are kept more closely in check.

Cats and Insects increase.

Several instances have been reported of local increase of insect pests as a direct result of the destruction of birds by cats. Mr. T. Bennett of Chicago writes that birds were abundant and his garden produced well, but new neighbors came in with cats, six of which now visit the garden regularly. Last summer, he says, half the birds were killed. This year hardly one is left, and many spring migrants have disappeared. He never knew before that there could be so many destructive insects in a square foot. "Bugs and worms" had to be fought on everything. Flowers and vegetables were poor and nearly a failure.[2]

Injury by Insect Pests.

Insect pests introduced from foreign countries added to native pests have become so destructive that, according to our best sources of information, the loss to agriculture and forestry from insect ravages in the United States exceeds a billion dollars

---

[1] The Bureau of Biological Survey of the United States Department of Agriculture has taken a preliminary bird census in the northeastern States, including those north of North Carolina and east of Kansas, and finds that farm land averages but one pair of birds to the acre. Professor Cooke, in reporting on this census, opines that the present bird population is "much less than it ought to be and much less than it would be if birds were given proper protection and encouragement," and he cites farms where the birds average 3 pairs to the acre, one having 4 pairs to the acre, and one section of 23 acres, thickly populated, where the birds average nearly 7 pairs to the acre. Where the birds were most carefully protected there were 13 pairs of birds nesting on half an acre. It is noteworthy that the numbers of domestic cats on this area are "below the average." Cooke, Wells W.: Bull. 187, U. S. Dept. of Agriculture, 1915, pp. 6–9.

[2] Bird-Lore, Vol. 12, March–April, 1910, pp. 79, 80.

annually. According to a conservative estimate made by Dr. H. T. Fernald of the Massachusetts Agricultural College, in 1901, insects were then costing the people of Massachusetts $4,400,000 annually. Using the same basis for estimation, we find that the annual loss now (1915) would reach nearly twice that amount, and it may exceed even that sum, as the expense of the fight against insects has increased in greater proportion than have the insects themselves. In 1890 Massachusetts appropriated $25,000 for the fight against the gypsy moth. Since then other foreign pests have appeared, including the brown-tail and leopard moths, the elm-leaf beetle and the San José scale, so that the money actually expended in one year by State and national governments, towns and cities, associations, etc., for the suppression of these insects in Massachusetts has reached the tremendous sum of $750,000 in one year (1913). Therefore it seems not improbable that all the insect pests of Massachusetts cost the people $9,000,000 in 1913. Dr. Fernald writes that he would not be surprised if the cost should prove to be at least as much as that. It is now well known that birds eat quantities of many of the most destructive insect pests, including the gypsy moth, the brown-tail moth, the elm-leaf beetle and the leopard moth. The last, which has destroyed many highly valued fruit and shade trees in Boston, Cambridge and other cities, makes no progress and does no appreciable damage in rural districts, where native birds are plentiful.

About fifty species of birds feed on the gypsy moth and the brown-tail moth. These birds must be protected and increased if possible. Instances have been recorded where flocks of cedar waxwings have freed many elms from the leaf beetle. Every bird that is useful in destroying all these insects is found on the list of the cats' victims.

Insect Pests eaten by Birds.

Following is a list of some of the most destructive insect pests that are eaten in great numbers by some birds that the cat commonly kills.

Insect.	Plants injured or destroyed by it.	Birds eating it.
Gypsy moth and brown-tail moth.	Fruit, shade and forest trees,	Cuckoos, robin, bluebird, jay, oriole, vireos and many others.
Codling moth,	Parent of the apple worm which injures the fruit.	Woodpeckers, chickadee and others.
Tent caterpillar,	An apple and cherry pest,	Cuckoos, jay, chickadee and many others.
Forest tent caterpillar,	Fruit, shade and forest trees,	Cuckoos, warblers, waxwing, oriole and many others.
Webworms,	Fruit, shade and forest trees,	Cuckoos, jay, chickadee and many others.
Army worms,	Grass, corn, etc.,	Robin, sparrows, bluebird, blackbirds and many others.
Cutworms,	Nearly all crops,	Robin, catbird, bluebird, blackbirds, sparrows and many others.
Cankerworms and other geometrid caterpillars.	Injure fruit and other trees,	Nearly all birds of orchard or woodland.
Cabbage worm,	Cabbages,	Song sparrow, chipping sparrow, towhee.
Beet worm,	Beets,	Chipping sparrow.
Colorado potato beetle,	Destroys the potato and egg plant.	Bobwhite, yellow-billed cuckoo, rose-breasted grosbeak.
Elm-leaf beetle,	Kills elms,	Cedar waxwing, vireos, etc.
May beetles and their young, the white grub.	Grass and garden plants,	Robin, blackbird, thrasher, catbird, towhee and others.
Rose beetle,	Roses and other plants,	Wood thrush, martin and others.
Cucumber beetle,	Destroys cucumber and squash plants.	Oriole, martin, phœbe, nighthawk, etc.
Weevils,	Fruit, clover, grain, peas, beans, etc.	Eaten by very many birds, bluebird, oriole, downy woodpecker, etc.
Click beetles and wireworms,	Roots of many garden plants,	Robin, sparrows, oriole, phœbe and many others.
Plant lice,	Plant life generally,	Warblers, chickadee, sparrows, thrushes and others.
Bark lice,	Fruit and other trees,	Nuthatches, chickadee, creepers.
Scale insects,	Fruit and other trees,	Chickadee, grosbeak, etc.
Grasshoppers and locusts,	Grass, grain and other crops,	Practically all birds.
Crickets,	Grass, grain, fruit, etc.,	Many ground birds.

This list might be extended almost indefinitely space permitting, but it is not enough that birds eat these insects; they must destroy large quantities of them or their services in checking the swarms of insect life never will be appreciable.

Number of Insects eaten by Birds.

Often in examining the contents of birds' stomachs, remains of so many insects are found in them that the number seems so incredible as to indicate that these fragments must have remained in the stomach for days; but experiments have shown that food passes the entire digestive tract of a small bird in from twenty minutes to an hour and a half, depending on the species and the kind of food, and that they require several or many full meals daily to keep up their high temperature, rapid circulation, quick respiration, rapid digestion and unusual muscular activity. Experiments have demonstrated, also, that many birds, particularly the young, consume more than their own weight of insect food daily, and that it is not unusual for a pair of birds and their young to dispose of from 300 to 1,000 insects a day. If they feed on minute or newly hatched insects, the number may be far greater. Dr. Brewer's calculation that a family of jays will consume a million caterpillars in a season may be an exaggeration, but it shows what an impression the study of this bird's food habits left on his mind. I have given much attention to this subject and have written more fully on it elsewhere.[1]

Various estimates regarding the number of insects killed by birds in different States have been made. Reed calculates that the birds of Massachusetts destroy 21,000 bushels of insects daily from May to September.[2] A Nebraska naturalist has estimated that the birds of that State eat 170 carloads of insects per day, and it has been calculated that the birds of New York destroy more than 3,000,000 bushels of noxious insects each season. These figures may be wide of the actual numbers but they are based on known facts.

Birds save Trees and Crops from Destruction.

I have noted many instances where birds have saved trees and crops from destruction by insects, and many where the destruction of birds has been followed by a great increase of insect pests.[3] In 1894, a year of insect abundance, I succeeded in protecting an orchard in Medford, by attracting birds, thereby securing the only full apple crop in town that year, while my nearest neighbor got a partial crop as a result of my experiment.[4] Baron Hans von Berlepsch kept his forest in fine condition by attracting and protecting birds on his large estate Seebach, in Angensalza, Thuringia, Ger., at a time when all the other trees of the countryside were stripped bare by caterpillars. The beneficial effect produced by the birds extended for a quarter of a mile beyond his boundaries. The baron does not tolerate a cat outside the buildings.[5]

Bobwhites have been more numerous on my place this summer (1915) than for many years. They have frequented the potato patch, and for the first time in years it has not been necessary to spray for potato beetles. I have recently received the crop

---

[1] Useful Birds and their Protection, published by Massachusetts State Board of Agriculture, 1907, pp. 41–63, 153, 154, 162.
[2] Reed, Chester A.: Introduction to the Bird Guide, 1905.
[3] Useful Birds and their Protection, published by the Massachusetts State Board of Agriculture, pp. 63, 76.
[4] Birds as Protectors of Orchards, annual report of the Massachusetts State Board of Agriculture, 1895, pp. 347–362.
[5] Heisemann, Martin: How to attract and protect Wild Birds, 1912, pp. 50, 51.

of one of these birds, sent me by Mr. Chas. P. Curtis of Boston. The bird was killed by a mowing machine in the field; but the crop contained 48 potato beetles and 250 weed seeds. Mr. James Henry Rice of Summerville, S. C., writes that by protecting bobwhites, and encouraging them to breed in and about his potato fields, he has secured practical immunity from the potato beetle. These examples are quite enough to show that birds in sufficient numbers may become important checks on injurious insects. It is difficult to compute the value of birds to agriculture, but Mr. Wm. R. Oates, State fish, game and forestry warden of Michigan, has placed the value of insectivorous and seed-eating birds of that State at $10,000,000 per year, and doubts if an equivalent could be secured in human labor for twice that amount.[1]

If we assume that a bird, during its normal lifetime, eats but 50,000 insects, each cat that kills 50 birds in a year saves an enormous host of insects, the number varying in each case with the potential length of life of the bird had it not been killed by the cat. A cat that kills only 10 birds annually protects a swarm of insects. It is fortunate that some few of the insects commonly eaten by birds feed on injurious insects, otherwise the destruction of birds by cats would be even more serious.

Inutility of the Cat.

No statement of the food of the cat would be complete without reference to an analysis of the stomach contents of a few hundred stray or feral cats taken in the open country. I have made no attempt to obtain such a collection for the obvious reason that a price offered for such stomachs might result in the destruction of many pet cats. The known facts, however, are sufficient to warrant the conclusion that the domestic cat, straying in the fields and woods, whether a pet, a vagabond or a wild, free dweller in the open, is a menace to wild life and a detriment to the general welfare. Doubtless, in its native wilderness this little feline was an essential part of the faunal life of the continent. It found abundant food, either in the forest, the jungle or in the open veldt, fulfilled its part in holding in check the swarming forms of smaller animal life, and its own carcass furnished food for the larger canines and felines that preyed upon it. When introduced into the New England fields, it became at once a disturbing foreign force, increasing beyond reasonable bounds, — a fruitful source of trouble. As most of its

---

[1] Biennial Report, Game, Fish and Forestry Department of Michigan, for 1913–14, p. 27.

enemies have become practically extinct in the greater part of New England, its increase is bounded only by the limit of its food supply and the activity of hunters and trappers, who have no pecuniary incentive to destroy it, as its fur is of trifling value.

While the cat is not indispensable in buildings, and while mice and rats may be held in check and locally exterminated without a cat, an efficient mouser and ratter will often do more to keep down the numbers of rats and mice than would the ordinary miller, grocer, farmer or householder if he had no cat. Unquestionably, then, selected cats are useful in the dwellings and granaries of man, as a check to the increase of small rodents, but when allowed to roam out of doors the species becomes a serious detriment to the agriculturist. Even if we take no account of the birds that it destroys, the balance would weigh against it, and when the results of its bird-killing habits are examined, it becomes a decided evil.

ANIMAL SUBSTITUTES FOR THE CAT.

Both before and since cats were first tamed other animals have been utilized to destroy rats and mice. Some have been tamed and domesticated, others have been kept in confinement except when in use, and still others have been merely tolerated. Snakes have been tolerated or utilized in buildings and dwellings as ratcatchers from time immemorial. The owl, weasel, stone marten, polecat, ferret, mongoose, skunk and dog have been made use of as ratcatchers. Weasels, as hereinbefore stated, are admitted to be far superior to cats, as they can follow both rats and mice into their holes, but, like the ferret, they must be kept in confinement or under control. It is said that rats and mice will not enter a building in which a weasel is kept, and that the coming of a weasel to a building will drive out all rodents that escape it. The ancients are believed to have used weasels and stone martens to rid buildings of rats, controlling them when at work by means of long chains, which allowed them to run into rat holes, but the most successful animal rat hunters of the present day are well-trained dogs and ferrets working together. The muzzled ferret drives out the rats and the dog catches them. Ferrets and dogs, however, must be trained, fed and accustomed to work together, and must be attended and assisted by their master. No dogs are better for this purpose than certain small terriers, particularly the fox terrier. Such dogs, working with ferrets and under the direction of their master, will kill enormous numbers of rats, and will practically exterminate them from any premises in a short time. Airedales can be trained to kill both cats and rats. Cats are preferred, however, by most people, particularly by the poor, because they may be had for the asking, or without asking, cost little or nothing to keep, care for themselves, hunt without aid, usually will not desert their home when given liberty, and make pretty and pleasing pets. Personally I prefer ratproofing and traps, but there are conditions under which cats or dogs and ferrets may be useful.

IS THE CAT A DISSEMINATOR OF DISEASE?

It has been regarded as a possibility that the germs of certain diseases may be carried in the mail and that the recipients of such mail may be infected. How much greater might be the chances of infection from the household pet going from the sick room to other rooms or dwellings!

Many writers on the cat include a long list of diseases to which the animal is subject, some of which are known to be deadly and contagious. Therefore, the questionnaire sent out from the office of the Massachusetts State Board of Agriculture contained the following question: —

Do you know of cases of contagious diseases carried to human beings by cats?

There were 222 negative replies and the rather surprising number of 67 affirmatives, reporting 17 diseases apparently transmitted by cats. The number of cases reported is much larger than this, as several correspondents noted more than one case. A majority of the physicians replying cited cases of infectious diseases transmitted by cats. This led to an investigation which shows that the cat is a rather neglected factor in sanitary science. Some physicians insist that cats shall be banished from the sick room or strictly quarantined, but their presence there is not generally considered dangerous.

Some sixty pages of evidence regarding the transmission of infection from cats to man was collected, mostly from medical sources. This to the layman looked convincing, but as much of it was of the character denominated by the courts as circumstantial, it was first somewhat condensed and then submitted to an authority on preventive medicine, who at once disposed of some of it as untrustworthy and regarded much of it as based on speculation, and as unconvincing to the careful scientific investigator.

It is undeniable that the cat may be affected by certain diseases and that it may transmit some infections, such as scarlet fever or smallpox, to man. But in the nature of the case much of the evidence is not such as would convince the bacteriologist, and probably some recent writers have inadvertently exaggerated in the popular prints the danger of infection from the cat.

Nevertheless, it will be conceded that as a carrier of disease, especially to children, no animal has greater opportunities. Any domesticated animal may act as a distributor of disease. Even fowls and pigeons have been accused of the offense; but the relations of the cat with mankind and with other domesticated animals and rodent pests are such as to suggest increased chances of spreading infection. It exceeds all other domesticated animals in numbers. It is less under control than any other. It is more generally allowed to enter sick rooms, sleeping apartments, kitchens, living rooms and places where food is kept, and is more likely to come in contact with milk. Its small size gives it an opportunity to creep into filthy places where most dogs cannot enter. Its habits of pawing over garbage and manure, and of rolling in dirt and clawing or pawing it, seem to suggest unpleasant possibilities, particularly as it comes commonly into close contact with the mouths and nostrils of children. The licking of its fur, by which infectious matter — peculiar to its own diseases — may be smeared over its whole body, may be weighed also in considering the likelihood of its spreading disease.

Dr. Caroline A. Osborne was the first to make a special effort to call public attention to the possible danger of infection by means of the cat, in a paper entitled "The Cat, A Neglected Factor in Sanitary Science."[1] This was followed by another paper entitled "The Cat and the Transmission of Disease," published in the "Chicago Medical Recorder" in May, 1912. In these papers Dr. Osborne maintains that science demonstrates that forms of animal life living with man may become infected with human disease organisms, and may transmit those organisms to man as well as to each other. The cat is the pet of small children, is handled, hugged and kissed by them, often becomes the playmate of a sick child, and is allowed to wander into the street where it meets other cats, or into other houses where it is fondled by other children.

Cohen says that domestic animals, especially house pets, and homeless cats and dogs probably are responsible for many cases in local quarantine.[2]

---

[1] Pedagogical Seminary, Vol. 14, No. 4, December, 1907, pp. 439-459.
[2] Cohen, Solomon Solis, editor: System of Physiological Therapeutics, Vol. 5, 1903, pp. 144, 340.

An editorial in the "New York Medical Record" for June, 1906, says: —

No one who has witnessed the enthusiasm with which children caress their pets can fail to realize the magnificent opportunity for infection offered in this. The doctor must in the interest of public health, see to it that no cat is allowed to enter a sick room.

The evidence at hand shows that cats have been accused or suspected of transmitting more than a score of infections to man or domestic animals. The diseases named range from scarlet fever, smallpox and bubonic plague to whooping cough, mumps and foot-and-mouth disease. Science already has acquitted the cat in some cases, and future investigation may either confirm or deny other allegations. There are some infections, however, regarding which the evidence seems conclusive.

Parasitic Diseases.

Cats are notoriously subject to a parasitic skin disease commonly known as ringworm, which is not uncommonly communicated to persons. Dr. James C. White of Boston asserts that he has known of many cases of ringworm carried to persons by cats. Dr. John B. May refers to an epidemic of ringworm in Waban, caused by a cat. Many others cases might be cited.

Cats may have external and internal parasites, some of which are or may be transmissible to man, of which space will not allow the enumeration here. Sand fleas, cat fleas, dog fleas, rat fleas or human fleas may be carried by cats. Those who care to know more of the internal and external parasites which cats may disseminate are referred to Dr. Osborne's papers hereinbefore cited, and the bibliography appended thereto.

Infections from Cats' Claws and Teeth.

Many painful and sometimes dangerous or even fatal inflictions are recorded as arising from the teeth or claws of cats, which they use freely against their human friends or enemies on the least provocation.

Tetanus or Lockjaw.

There is no more fatal or awful disease than this. Unless tetanus antitoxin is injected early there is practically no hope for recovery. Many cats live about barns and stables. In burying their own excreta their claws often come in contact with horse manure as well as dirt, both of which may be infected with the germs of tetanus, which often swarm in the former, but only one case of lockjaw from a cat scratch has been reported to me.

Rabies or Hydrophobia.

All authorities agree with Pasteur that the cat is a medium through which this disease increases in virulence for mankind. The bite of a mad cat, therefore, is even more dangerous than that of a mad dog.

Rabies has been noted in Germany since 1809 among cats, and the evidence seems to indicate that it was acquired from foxes. A fox attacking poultry had an encounter with a cat which, being bitten, later bit a servant girl who died of hydrophobia. In those days no remedy was known and fatalities were numerous. The disease became epidemic among both wild and tame cats. It spread widely, raging until 1827, and extending to Norway, Denmark, England and elsewhere, including among its victims dogs and wolves.[1] Many people were bitten.

In recent times the infection has been considered rare among cats, but public attention has been called to this danger by the recent death of little Grace Polhemus, of 372 Monroe Street, Brooklyn, N. Y., which occurred in spite of the Pasteur treatment. In this case the evidence of the cause and nature of the infection and death of the child are conclusive. Thirteen years old and in perfect health, she was playing in the front yard of her home when she stooped to pet a stray cat, which bit her on the right wrist. Letters from Dr. Albert Thunig, Brooklyn (who was associated with Dr. Vosseler of Brooklyn in the care of the case), and Dr. F. T. Fielder, assistant director in the vaccine laboratory of the health department of New York City, contain the following evidence: —

(1) The child was bitten by a stray cat, Oct. 18, 1913, and treated by a physician (wound sterilized with iodine) within a few minutes. (2) The cat was captured, placed in charge of the health department, its brain examined after death at the research laboratory, and negri bodies found, proving that it had rabies. (3) The Pasteur treatment supplied by the department of health was administered to the patient by a physician for twenty-one days. (4) There was no other bite or infection between this treatment and the time of the development of the disease. (5) Characteristic symptoms of rabies began to appear November 7, and as the symptoms progressed, it was evidently a "classical, clinical" case of rabies. Death occurred November

---

[1] Fleming, George: Animal Plagues: their History, Nature and Prevention, Vol. 2, 1882, pp. 15, 16, 74–77, 80, 89–91, 95, 99.

13. (6) The brain of the patient was examined at the research laboratory, department of health, and negri bodies were found. (7) Guinea pigs inoculated with cultures from this brain contracted rabies two weeks after inoculation, thus confirming the diagnosis of rabies as the cause of the girl's death.

Dr. Fielder volunteers the information that the research laboratory of the health department examines a considerable number of cat brains yearly, as many people are bitten each year, and that in 1913, 14 out of 46 cats examined proved to be rabid. About 50 people in New York are obliged to take the Pasteur treatment each year "because of bites by rabid cats, or by stray cats possibly rabid which escape and so cannot be examined."

Dr. John B. Huber asserts that in the last six months of 1914, 42 persons bitten by cats received Pasteur treatment. The cats that bit 33 of these persons were examined in the New York City laboratory and proved to be rabid. Mr. Harold K. Decker of West New Brighton, N. Y., writes that a mad cat bit several people in that neighborhood in 1914; it bit a dog which also became mad and bit other dogs and cats. The people bitten were saved by the Pasteur treatment.

Rabies among cats has a long history. Fleming, an authority on this infection, says that dogs and cats "hold first place in the scale of susceptibility."[1] He reports or cites the loss of a large number of human lives by hydrophobia induced by the bites of rabid cats.[2]

Septicæmia or "Blood Poisoning."

The following list shows a number of more or less serious injuries resulting from the bites and scratches of cats, as reported by my correspondents: —

Injury.	Number reporting it.
Serious bites (1 fatal),	28
Serious scratches,	27
Blood poisoning from bites,	9[3]
Blood poisoning from scratches,	7[4]
Fatal,	1
Damage to eyes,	7
Loss of eye,	1
Corneal and other ulcerations of eyes,	1

[1] Fleming, George: Rabies and Hydrophobia, 1872, p. 92.
[2] Ibid., pp. 47, 54, 55, 60, 64, 147, 246.
[3] One caused loss of use of arm for two months; another caused loss of a part of one hand.
[4] One caused loss of two fingers; one caused death of infant.

Perhaps there is no conclusive evidence in any of these cases that infection of septicæmia came directly from the teeth or claws of the cat, as the wounds caused by the cat might have become infected from some other source after they were inflicted, and similar results might arise from the scratch of a nail or a piece of tin, but the claws or teeth may have been the medium of infection, and such cases are not very rare.

A perusal of the above should cause parents to consider whether cats or kittens are likely to be safe playmates for their children, or whether harmless creatures like rabbits are not preferable.

As a precaution against possible infection tramp cats should be eliminated, sick ones quarantined and all cats should be kept away from the common sources of infection, especially from all people ill with transmissible diseases.

Boards of health of towns and cities cannot ignore the cat as a possible agent in carrying disease infection. Medical men are now banishing cats from hospitals and other institutions. The following letter from the commandant of the naval training station at Newport, R. I., explains itself: —

Replying to your letter of July 23, inquiring relative to the destruction of cats at this station, you are informed that all stray cats found on this station were a short time ago disposed of. Every effort is made here to prevent possible contagion to 2,000 young men, and this is one of the preventive measures.

Very truly yours,
ROGER WELLS,
Captain, U. S. Navy, Commanding.

MEANS OF CONTROLLING THE CAT.

If ownerless cats were eliminated and owned cats confined like other domestic animals, or limited in their movements to buildings or enclosures of their owners, the cat evil would be minimized. Even if the cat could be brought to obey a master and so be kept under control, like the dog, the trouble would not be so acute. The cat then could be utilized more in killing rats and mice and prevented from destroying birds; but the moment the average cat in the country gets away from the house it becomes practically a wild animal and beyond control, except by means of a shotgun or rifle. A well-trained dog will come at call, but most cats are not trained to obey any call, except that of an empty stomach.

As the cat is not a necessity, many people do not keep one. I have not kept a cat in my house for years. Whenever rats or

—mice get in we catch them immediately. I never have had a rat or a mouse in my summer camp, where no cats are allowed, but in the farmhouse near by, where two cats are kept, rats come and go, and in the barn and outbuildings, which the cats frequent, rats always exist in numbers, although rarely seen. I never use poison in my buildings. Ratproofing and traps properly used will free any dwelling house of rats and mice. Readers who do not know how are referred to Economic Biology Bulletin No. 1, "Rats and Rat Riddance," which may be procured by applying to the secretary of the Massachusetts State Board of Agriculture, Room 136, State House, Boston. Some catless people have little success with traps and are overrun with rats and mice. This happens because they do not know how to handle the rat problem, or have not time, skill, industry or persistence enough to outwit the rats. Others who have no cats have less trouble with rats than their neighbors who keep many cats.

Inquiry among correspondents who keep no cats elicited the reasons why they do without them, which fall under the following heads: (1) danger to children from bites, scratches and disease; (2) cats kill birds; (3) cats kill chickens; (4) antipathy for cats; (5) cats do more harm than good; (6) cats are unclean and make too much trouble.

Those who do not keep cats have not solved the cat problem, however, as many of them complain that their premises are overrun by neighbors' cats or stray cats, and that birds and chickens are killed by them. Nine complain of the destruction of young trees by cats' claws, 39 of damage to gardens by trampling and scratching in them, and 179 of disturbance by caterwauling.

Catproof Fence.

A catproof fence may be made by first setting up a chicken wire fence six feet high and attaching to the tops of the posts slim upright poles from which a fine fish seine is hung with its lower edge fastened to the top of the wire fence, thus making a barrier at least nine feet high. The fish net hangs so loosely from the slim poles that it gives beneath the weight of the cat and baffles the animal completely. The bottom of the fence should fit the ground closely, and there should be no trees near, on the limbs of which cats can climb and then drop inside. A fruit garden enclosed by such a fence is likely to become a paradise for birds, but it may become a playground for rats as well, and measures to kill them may be necessary.

The reasons why people keep cats are given by cat owners as follows: (1) as companions and pets; (2) to catch rats, mice and other rodents; (3) to catch birds and game for their owners; (4) to catch mice and rabbits to protect orchards; (5) to keep birds away from strawberries.

The keeping of cats as companions or pets, however important it may be, is a matter of sentiment and does not come within the scope of this paper, except as it tends to increase the market value of the cat. Many cats are carefully housed, confined and bred for exhibition at cat shows, and some of them sell for high prices, but we have the testimony of some cat breeders that most of these high-bred cats have little if any desire to catch rats or mice. Angora cats are said to let birds alone, but I have evidence from several observers proving that some Angora cats are very destructive to birds.

People who keep cats which are trained to bring in birds and game have no right to the possession of birds or game protected by law. They are law breakers and should be treated as such.

Farmers who feed grain to cattle, horses, pigs and fowls often feel that they must keep cats to catch rats and mice in their barns and poultry houses, as they find it less troublesome and expensive to keep a few cats that are practically self-supporting than to attempt to catch or kill rats. Many farmers see only the good that cats do as ratcatchers, and do not realize how much they may lose indirectly through the killing of insectivorous birds by cats. All who raise chickens desire to protect them against cats. Many cat lovers are bird lovers also, and many people who keep cats as pets wish to prevent them from killing birds. In response to many inquiries I have received much advice regarding these matters. The replies may be summarized as follows: —

Method recommended.	Number recommending it.
Kill the cat,	175
Confine the cat,	63
Feed the cat well,	54
Feed the cat raw meat,	4
Feed the cat no raw meat,	5
Keep the cat on leash,	2
Bell the cat,	30
Use care in placing food for birds,	7
Bird-boxes on iron pipe,	2
Cat guards on trees and nest-box poles,	22
Barbed wire on trees,	6
Thorny shrubs or vines to keep cat out of grounds or away from bird-houses,	1
Deep nesting boxes,	3
Nesting boxes placed high,	3
Keep only light-colored cats,	1
Chicken wire about food tables,	1
Air gun, stones, tin cans, torpedoes, etc.,	4
Electrocution,	2
Dogs,	3

Killing the Guilty Cat.

The method recommended by 175 observers, "Kill the cat," is a sure and safe one. This applies to both bird-killing and chicken-killing cats, although it is easier to teach a cat not to molest chickens than to teach it to let wild birds alone. Poultrymen almost always find that when a cat once gets a taste of chicken, the only safety lies in killing the cat, and the main reason that so few farmers' cats kill chickens is that the chicken-killing cat is very short lived, and has little chance to transmit its bad tendencies to offspring. Wild or stray cats, village and city cats, and not farm cats, are the chief chicken killers. If every bird-killing cat were killed, and those that give their attention mainly to rats were kept, we would have fewer cats, but the survivors and their progeny would be more useful and much less harmful than most cats now are. It is well known that many cats specialize. Some take to hunting rats and mice and rarely look at birds in the trees; others hunt birds mainly and trouble rats and mice very little; others hunt everything from insects to cock pheasants; still others hunt rabbits and game, and some rarely hunt at all. The useful and nearly harmless cat possibly might be produced by selection and breeding. A rat-hunting female cat, if allowed to nurse and raise her own kittens, usually rears some good ratters.

Confining or Tethering the Cat.

A good ratter when confined in a building with rats and mice will devote its attention to them. A cat that will not do this is worthless except as a pet or an exhibit in a cat show. During spring and summer, when birds are nesting and breeding, cats may be confined in buildings or cages. Let no one think it cruel to confine a cat. Of course, one unused to being deprived of its liberty is likely, if shut up, to set up a piteous mewing, but cats brought up in narrow quarters live happily, especially if they have mice and perchance rats to give zest to life. Many cats live most of their lives in cages, while many others are kept in buildings that they are not allowed to leave. If brought up in such quarters they are cheerful and contented. Miss Repplier writes as follows of cats in confinement: —

As a fact, imprisonment has scant terrors for the cat. It accords too well with her serene and contemplative disposition. Restless wanderer though she appears, and true lover of liberty though she is, and has ever been, she can yet live her life with tranquil enjoyment in a ship, on the seventh floor of an apartment house, in a granary which she is never permitted to leave, or in London's Tower. There were probably many French cats who passed their days meditatively in the Bastile, content to be immured with their masters, and accepting like philosophers the restraints and the indulgences of that ill-omened but singularly comfortable fortress. "Stone walls do not a prison make" for a creature whose independence of character remains untouched by the sternest and narrowest of environments. Rather perhaps does she feel herself a captive when surrounded too strenuously by the doting and troublesome affection of mortals, who cannot be made to understand or to respect her deep inviolable reserve.[1]

Dr. Burt G. Wilder of Brookline, who is fond of both birds and cats, proposes the following plan, which he carries out with his own cat in summer at Siasconset, and with modifications elsewhere at other seasons: (1) Only one adult cat to a family, an additional one if there is a barn or stable, each kept in its own place, and superfluous kittens promptly destroyed. (2) The cat to be fed regularly and before the family meals instead of after, and in the meantime prowling about and getting under the cook's feet or into the food, before or during meals. Feeding to be attended to by or delegated to one person, not left to chance. Scraps from previous family meal may be provided. (3) All cats to be confined during the night and fed before they are released in the morning. If properly trained they will defer attending to the calls of nature until released. If not, provide a pan with sawdust or dry earth resting on a large paper. (He says that his cat loafs or sleeps most of the day outdoors and never has killed a bird. Other well-fed cats have killed birds, but confining nights and feeding early may be helpful.) (4) All cats to be licensed; unlicensed cats to be killed, by shooting, if wild. This opens the much discussed question of cat legislation, which is considered on pages 97–100.

A cat may be tethered to an overhead wire in pleasant weather by means of a line and a snap hook. This gives outdoor conditions, allows the cat to exercise by moving back and forth, and probably will prevent it from catching birds, except possibly such young as may flutter in its way. There should be a stop near each end of the wire so that the cat cannot climb or become entangled. Both these expedients are feasible, and many cats now are kept through the summer in confinement, or on a leash in fine weather. The large cat shown in the photograph, owned by Mr. Bardwell Gladwin of Plainville, Conn., is tethered in this manner because of his fondness for chickens. He has been thus treated every summer for five years, and Mrs. Louise G. Lusk

---

[1] Repplier, Agnes: The Fireside Sphinx, 1901, pp. 99, 100.

says that he thrives and seems to regard his leash as a high honor. High-bred cats kept for breeding purposes necessarily are kept in confinement most of the time.

Keeping the Cat Indoors at Night.

Most important of all, the cat should be kept in the house or some building, cage or pen at night. Cats which hunt outdoors at night contract colds and diseases, and destroy more birds and game and fewer house rats and mice than at any other time. About 90 per cent of the cats are allowed to roam at night. The mother bird is slain on her nest by the unseen marauder or the young are taken when they first begin to stir at early dawn.

Feeding the Cat.

A well-fed cat must have meat, as that is the natural food of the species. Probably cats that are fed meat and given water are less likely to engage in an active hunt for birds and more likely to stay at home and lie quietly in wait for rats and mice than those that are poorly fed and have to find their own meat and drink. A little milk once or twice a day is not good or sufficient food for a cat. Cat lovers tell us that if we wish our cats to be good mousers we must feed them well, as they cannot stand watch long on an empty stomach, but they tell us also that if well fed they will not catch birds. Nevertheless, I have known cats, excellent mousers and ratters, rarely fed by their owners, and I have many reports of cats well fed and well cared for which spent a great part of their time in hunting and killing birds that they never ate. On the other hand, it may be possible to feed a cat so much meat that it will not hunt. The owner of a fertilizer factory, where dead horses were received continually, said that both rats and cats, glutted with meat, fraternized about the boilers on cold winter nights, and that the cats never troubled the rats; but experience goes to show that a bird-killing cat, like a man-killing lion or tiger, has acquired a practically incurable habit, and while overfeeding may check the habit in some, it seems to have no effect on others.

Belling the Cat.

The experiment of putting a collar and bell on a cat to prevent it from catching birds has been recommended by many people who have never tried it and by some few who have, but the most common experience seems to be that a cat which is skillful enough to creep upon a bird, is expert enough to keep the bell from ringing until the final spring. Belled cats catch birds, rats and mice and all forms of wild life; although the bell may save a few birds in some cases, it never saves helpless young. Mr. Niel Morrow Ladd of Greenwich, Conn., records the fact that a sleek, fat Angora cat, although burdened with 6 bells, brought in during one nesting season 32 birds and in the next 28, none of which it ate.[1] This cat is shown on Plate VI. in the act of killing a young catbird.

Cat Guards.

Most of the devices for protecting the nests of birds are useful against the cat only when nests are on isolated trees or in boxes on poles. Such devices will not protect nests on the ground in shrubbery or in woods. In such cases a tract of land may be surrounded with a very high, thick, thorny, and impenetrable hedge or a catproof fence. Nesting boxes on the perpendicular walls of buildings are inaccessible to cats, and those on tall slim poles are not often troubled by them. Nest boxes hung by wires have been recommended.

The plan proposed by Raspail, by which the nests both on the ground and in trees are surrounded and covered by a wire netting, to keep the cat away (see Plate XVI), allowing the bird to slip in through the meshes of the top, has been successfully used both here and abroad, but is expensive and is useless unless the nest is protected before the cat finds it. It is easier and less expensive to cage the cat rather than the nest, but the wire netting may protect the nest from wandering cats.

To puzzle cats.

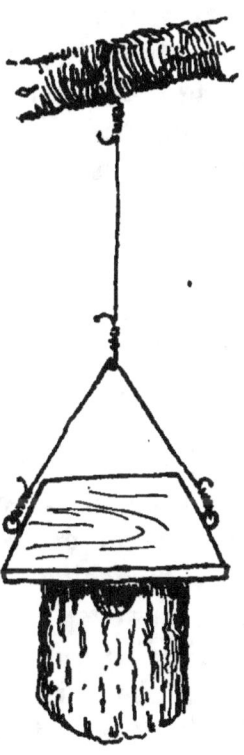
Difficult for pussy.

It is well known that cats are very sensitive, and that they are fond of catnip and other aromatic plants; also they detest certain odorous plants and substances. Housewives formerly tied slips of rue under the wings of chicks to protect them from cats. The odor of orange peel is said to disgust cats. In England cats once were singed to keep them at home. Hence the old

---

[1] Ladd, Niel Morrow: How to attract Wild Birds about the Home, 1915, p. 35.

saying about a singed cat. Chaucer has immortalized the practice in verse. It was believed that the cat was vain of its appearance, and that if the fur were well singed, shame would keep the creature at home. The Dundee (Scotland) "Advertiser" states that the French National Society of Acclimatization has taken up this cause of the destruction of game and birds, and has tried to find a remedy for it. "The society now informs us in its bulletin," says the "Advertiser," "that in order to keep the cats away from a bird's nest we have only to place a cloth or rag saturated with 'animal empyreumatic oil' in the bush or on the trunk of the tree where the nest is situated." Cats have an unconquerable repulsion for the smell of this oil. One correspondent having caught a mouse in a trap rubbed it over with empyreumatic oil and then let it go in the presence of his cat. The cat took no notice of the mouse. Whether the odor had been caught by the other mice in the house, or whether the cat kept a disagreeable reminder of the experience, he absolutely gave up chasing the mice which swarmed in the house. This method is worth a trial.[1] For additional cat guards see Plate XIX.

Keep only White Cats.

This suggestion, given by one observer, is good, as a white cat may find it difficult to catch full-grown birds in the daytime, *but the color will not save the young birds in the nests or those learning to fly.*

Air Guns, Torpedoes, Etc.

There is nothing more effective in frightening a trespassing cat than a well-directed shot from an air gun, a large torpedo thrown and exploded close by it, a tin pan thrown so as to clatter, or a drenching from a hose. These rather cruel expedients may not, however, prevent the same cat from returning at night and marauding at will. Mr. John Gould of Aurora, O., says that if a cat is shot with a charge of salt it will avoid the place ever after, but that is torture.

Electrocution.

This has been practiced on marauding cats by running heavily charged wires about the tops of hen pens or pheasant pens. It is too dangerous and expensive for general use.

Dogs.

A large, active, fearless dog may be trained to drive cats off premises, to tree them, or even to kill them, but must be on watch night and day, and may, meantime, eat eggs or molest some birds.

---

[1] Sixth annual report of the Massachusetts State Ornithologist, annual report State Board of Agriculture, 1913, p. 267.

Training the Cat not to catch Birds.

Weir says that cats may be trained to respect the lives of birds and other wild animals.[1] De Voogt says that the bird-killing cat may be easily corrected by "taking a bird in the hand and making it peck the cat's nose."[2] This might succeed with cage-birds.

I have never seen a cat that I felt sure would not catch a bird if given a good chance, except one that was blind, but I have been assured by people in whom I have every confidence that they believe that their cats never caught a bird, or that they have been taught not to catch them. Nevertheless, in some cases these good cats have been seen by neighbors in the act of catching birds.

Mrs. Elizabeth B. Davenport of Brattleboro, Vt., writes that she has taught cats to let birds alone, but that not one person in a hundred would have the patience to do it. The first one so taught was never allowed to keep a bird that he caught, and if he evaded her the hose was used. He was punished lightly if he went near birds, and was kept constantly in view when out of doors. The second season he ceased to watch them. A lady writes that she had a cat which absolutely would not catch birds. The birds seemed to have no fear of this cat, and sparrows dressed their feathers unafraid while it rubbed against the bush just below them. A few others make similar statements about their cats. Mr. C. J. Maynard of Newtonville, an experienced naturalist and a competent observer, says that he has two cats that never kill birds. *He taught them as kittens* to let birds alone by feeding them well and gradually accustoming them to seeing birds near, beginning with bird skins or mounted birds. This is a method, however, which cannot be practiced by all.

Correspondents report on this matter as follows:—

Know of a cat that will not catch birds,	70
Believe cats cannot be taught not to catch birds,	305
Believe cats can be taught not to catch birds,	62
By whipping,	37
By scolding,	8
Tying bird to collar or around neck,	9
Taking bird away from cat,	14
Drenching cat with water,	1
Pepper on dead bird,	2
Pepper and kerosene on dead bird,	1

---

[1] Weir, Harrison: Our Cats and All about Them, 1889, p. 12.
[2] Burkett, Chas. Wm., editor: Our Domesticated Animals. Translated from the French of Gos. De Voogt. 1907, p. 81.

I have had no success with any of these methods, and have known all to fail except that of putting pepper and kerosene on the dead bird. Many correspondents express the belief that many people who believe that they have taught their cats not to kill birds have merely taught them not to bring the birds in, but to catch them in the fields and eat them under some building, or to leave them where killed. Dr. Anne E. Perkins, writes that she used to be very fond of cats, and can speak from years of experience, both with her own beloved pets and with others. She asserts that much pains was taken to break them of bird killing, but after they had been punished they did not bring the birds in sight as they did with mice, etc., but many a heap of feathers was found. Others report similar experiences.

In 1914 a female cat took up her abode on my farm. She was believed not to kill birds, having been taught (?) by whippings when a kitten. For two months there seemed to be no evidence to convict her of bird killing, although I found a nest destroyed in one place and remains of young robins in another. Then she was seen with a bird, and later with another. A week later I found her with a live blackpoll warbler, and as I approached I heard her teeth crunch its tender bones, which prevented all chance of rescue. We tied the dead bird firmly about her neck, but she took to the woods, and in half an hour she had clawed it off and probably had eaten it. If the bird had been sewn in canvas or duck the expedient might have been more effectual. The plan of securing the bird firmly about the cat's neck and leaving it there until it "wears" off is said to be very effective. Red pepper may sometimes prevent a cat from eating a bird, but in several cases reported to me the cats ate the birds, red pepper and all. Kerosene probably is more effective, but all these devices may fail to prevent the cat from *killing*, as no one can possibly *know* how many birds his cat kills unless he keeps the cat shut in at night and under watch all day. Any one who succeeds in awakening the regard and affection of a cat may restrain it by constant watchfulness and words of displeasure or light blows upon the body (never on the head), but few people have the time or patience for this.

Some cats may be taught not to kill caged birds. *Kittens* in bird stores are so trained by means of red-hot knitting needles placed in front of a cage, when they first attempt to catch the birds, or by red pepper and kerosene on a dead bird, which teaches them to leave it alone.

To prevent Cats killing Chickens.

Chickens kept in coops covered with small meshed wire netting are safe from cats, but chicks often are stunted by such confinement.

Kittens brought up in the chicken yard or henhouse rarely kill chicks. Where a kitten shows a chicken-killing tendency it may sometimes be "cured" by shutting it in a small yard with a spirited hen and her brood. The hen will administer the treatment. If the offender is a grown cat the plan suggested by Mr. Wm. Lawlor of Needham may be better, otherwise the hen may come out second best. Mr. Lawlor suggests tying a cat up in a bag with its head out and dropping it in the yard with a savage old "setting hen." This would deprive the cat of some of its natural weapons of offence, but the bag should be a strong one. I have seen a cat confined in a pillow-case tear it open in a few seconds. Some poultrymen tie a chicken killed by a cat around the cat's neck and leave it there until it becomes offensive. Several persons report good results from this method.

LEGISLATION FOR THE CONTROL OF THE CAT.

We now legislate to protect birds, but place no limit on the increase and activities of their most destructive natural enemy. A man is liable to a fine if he kills a bird, but he may with impunity keep any number of cats to kill birds and bring them to him, although he has no legal right to possess or use birds so caught. Many people believe that a statute should be enacted to limit the numbers and activities of cats, and that such a law should provide responsible officers to kill surplus cats, and should furnish the money to pay them for their services.

Mr. Winthrop Packard of Boston proposes the following plan for cat legislation: (1) License every cat and make the fees — male, $1; female, $2. (2) Make the license operative as a protection to the cat only while it remains on the owner's premises. (3) Make it a misdemeanor punishable by fine to own or harbor an unlicensed cat. (4) Require owners of licensed cats to keep a collar on each such cat, bearing on a suitable tag or plate the number of the license and the name of the owner. (5) Require duly authorized officials to kill unlicensed cats in a humane manner. (6) Pay such officials out of the money obtained for cat licenses.

These regulations would be excellent from the standpoint of the cat breeder, most bird protectionists or that of the public health authorities. Strong objections to them come, however, from many people.

1. Many cat keepers object on account of the tax. The strongest objections come from those who keep the largest number of cats. No one likes to be taxed. The cost of living in this country is high, and most farmers, many of whom believe that they pay more than their share of taxes, because their property is all visible and cannot be concealed, oppose the tax strenuously. Nevertheless, it would benefit the farmers more than any other class, as the destruction of stray and unlicensed cats would save birds and chickens enough to far more than pay the tax. Friends of this legislation argue that a male cat which is not worth at least one dollar to the owner as a rat and mouse killer, or as a pet and companion, ought to be humanely executed, and the female cat, which usually is a better ratter than the male, will, if worth keeping at all, easily save the farmer far more than her license fee by destroying rats and mice. If only the useful and valuable cats could be kept, and the worthless ones destroyed, the aggregate saving of birds would be enormous.

2. Most farmers object to being obliged to keep their cats at home, because it is difficult, if not impossible, to do so and at the same time give them such freedom as they need in catching rats and mice on the farm. The advocates of these regulations say that this difficulty may be met by keeping cats in the buildings as much as possible, feeding them well and breeding from those that manifest little desire to roam. Enforcement of the law would tend gradually to eliminate the wandering and stray cats, and leave only the stay-at-homes, which in most cases are most desirable.

3. Only lawbreakers will object to the fine for harboring and keeping an unlicensed cat.

4. Many people object to putting a collar on a cat because of the belief that the animal may be hung by it, while climbing trees, and cite cases where cats have been so hung, and many cases where collars have been put on loosely and have come off. But the proponents of the legislation reply that while there may be danger of cats becoming entangled and strangled by the wearing of loose collars, which may be caught in the branches of trees, there is practically no danger if the collar is fitted snugly to the neck of the animal, and they point to the many cat owners who keep such collars on their cats, and to cats that have worn such collars for years without accident. Mr. Wilfrid Wheeler, secretary of the Massachusetts State Board of Agriculture, asserts that he kept a collar on a cat seven years, until it came apart

and dropped off, but it never troubled the cat in the least. This objection to the collar might be met in many cases by tethering wandering or tree-climbing cats when out of doors.

5. Some people object to a cat license on the ground that the stray animals would not be humanely caught and killed, and that it would be impossible to catch them all. The proponents of the legislation reply that this work might be left to the Animal Rescue League in Greater Boston, as well as in other cities, wherever and whenever the league succeeds in establishing branches, and that as the laws relating to cruelty to animals are strict, there need be no unnecessary cruelty allowed. Also they assert that the great number of cats annually destroyed in Boston and New York by humane associations is sufficient proof that stray cats in the cities can be caught by experienced persons. In the country, expert men would have far less trouble to get cats that run wild than in the cities, where shooting and trapping must necessarily be limited.

The cat license is not a new idea. It was first advanced by humane societies and cat lovers as a means of protection to cats. The licensed dog is regarded as property, and as such has some rights, while the status of the cat is very precarious. It was argued that if cats were licensed they would be entitled to be regarded as the property of their owners, and could not be seized or killed with impunity.

Gordon Stables, cat lover, writing in 1876, says: "I should like to see a tax imposed upon all cats, and a home for lost cats precisely on the same principles as the home for lost and starving dogs."[1]

Miss Helen M. Winslow, cat lover, writing in 1900, advocates a cat license in the following words: "If our municipalities would make a cat license obligatory, just as most of them have ordained a dog law, placing even a small yearly tax on every cat, and providing for the merciful disposition of all vagrant, homeless ones, not only would there be fewer gaunt, half-starved prowlers to steal chickens and pigeons, but the common house cat would rise in value and receive better care."[2]

Recently such legislation has been proposed in many States, and we find many cat lovers in opposition. The leader in the movement to tax cats was Mr. Albert H. Pratt, president of the Burroughs Nature Club of New York, and there was much agitation on the subject in legislatures and municipal governments, but so far as I know, the only place in America, where

---

[1] Stables, Gordon: The Domestic Cat, 1876, p. 157.
[2] Winslow, Helen M.: Concerning Cats, My Own and Some Others, 1900, p. 263.

the cat license is operative (1915) is St. Petersburg, Fla., and Montclair, N. J., has an ordinance under which all owned cats must wear distinguishing tags or collars, and cats not so marked are humanely destroyed. Iowa has a State law under which cats might be taxed, but this opportunity has not yet been utilized. Certain bird lovers oppose the proposed law on the ground that it gives the cats more protection than they now have. Any tax always is unpopular. Nevertheless, there seems no other way to reduce the cat population within reasonable bounds by legislation, and at the same time provide for the enforcement of the law. No one is competent to pass upon the advisability or probable effect of cat license legislation until it has been tried and perfected in the light of experience. No doubt such trial will be made.

METHODS OF TAKING AND KILLING STRAY OR FERAL CATS.

Most cats may be taken easily in a box trap baited with catnip tied up in a cloth, or with fish. Cats are inordinately fond of fish, and are strangely attracted by the scent of catnip. Sometimes in summer when birds are plentiful cats will not come to a trap baited with fish. Catnip is then the best bait. The trap should be large enough to contain any cat and so made that

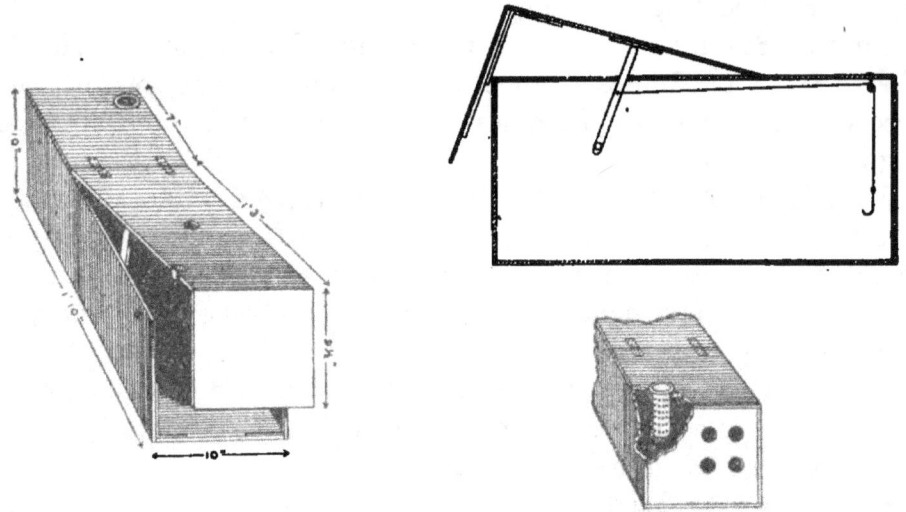

Mr. Huntington Smith's humane trap, with details.

the door or lid latches when it is sprung. A hole may be left open at the back and as the cat will come to this hole for air, it may be shot in the brain with a small rifle or pistol. Such a death is sudden, comes without warning, and as it is absolutely painless it is the most humane death possible.

A humane trap has been devised by Mr. Huntington Smith, of the Animal Rescue League, 51 Carver Street, Boston. It is 22 inches long, 10 wide and 9½ high. The bait is suspended on a hook that releases a cover, which drops and locks but does not shut tight, and therefore never even pinches the cat's tail.

The opening under the drop lets in air, which passes out through holes at the other end of the trap, thus giving ventilation. There is a receptacle for a sponge, into which chloroform may be poured, not coming in contact with the cat.

There is a trap on the market that chloroforms the cat as soon as it is caught. This is a humane trap but gives no chance for discrimination. It may chloroform the wrong cat.

The stop-thief trap is said to be humane because it garrotes the cat and quickly shuts off sensation. It is set at the entrance of a hole or passage, or at the mouth of some receptacle, so that the cat must reach through the trap to get the catnip with which it is baited. No. 3 is the size commonly used. Stables says, "Never drown a cat. If there is any one that can be trusted, who knows how to use a gun, by all means have her shot. It is over in a moment. The next best plan is to administer morphia. Don't grudge her a good dose — five or even ten grains. Cats are wonderfully tenacious of life, but they can't stand that. Make the morphia into a pill, with a little of the extract of liquorice, and force it down the throat. The cat will soon die and will not suffer.[1]"

Stop-thief trap.

Trapped cats may be chloroformed in a box trap by inserting through the hole in the back a sponge saturated with chloroform, closing the hole and covering the trap with a heavy blanket. Occasionally a stray cat may be too wary to enter a trap. Some that are suspicious of a trap closed at one end will enter one open at both ends. Any cat may be caught by burying or covering several smoked or carefully cleaned steel traps and scattering bait among them, but it is much less cruel to track the cat with dogs, and when it takes to a tree it may be shot through the brain with precision and certainty, suffering no pain. A crack shot with a rifle will make sure to bring down the game at the first shot. Others should use a chokebarreled shotgun, with a heavy charge of powder and shot not smaller than No. 4; BB shot might be better at long range. It is useless to shoot small shot at cats except at very close range. The head shot is the only sure and instantly fatal one. If shot through the body, the cat may live for some time.

---

[1] Stables, Gordon: The Domestic Cat, 1876, p. 88.

The trail should be taken at daylight while it is still fresh. On the first light snow of winter, the hunter does not need dogs, but starting early in the morning he follows the trail afoot, and kills every woods cat that he trails. In this way a tract of woodland may be speedily cleared of wild or half-wild cats, but the next winter others may be tracked and killed there. In the village or city a person whose personality attracts cats can pick them up rapidly. A kind word from such a person or a little attractive food will entice many a wandering and starving cat. On the other hand, when cats have been persecuted they are like the wicked that "flee when no man pursueth," and then one must resort to the gun or trap. Any man who can trap the fox or even the wary, experienced rat, can take any cat that lives. Recently a pet cat taken in a trap was drenched with water and liberated, but was caught again in the same trap within twenty-four hours.

LEGAL RIGHTS OF THE CAT.

During the past century cat lovers have made many attempts to prove that their pets are entitled to some rights under the law, but English law seems to find little merit in their claims. An articled clerk, writing to the "London Standard," says: —

It is clearly laid down in "Addison on Torts" that a person is not justified in killing his neighbor's cat or dog which he finds on his land, unless the animal is in the act of doing some injurious act which can be prevented by its slaughter. If a person sets on his land a trap for foxes, and baits it with such strong-smelling meat as to attract his neighbor's dog or cat on to his land to the trap, and such animal is injured or killed, he is liable for the cat, though he had no such intention and though the animal ought not to have been on his land.

The French courts have given the cat owner no damages in such or similar cases. The local magistrate of Fontainebleau heard a case in which a man, annoyed by neighboring cats, kept traps in his garden and caught fifteen. The neighbors combined to bring him to justice. The judge decided in favor of the neighbors, but in a higher correctional tribunal the decision was reversed.[1] In some European countries cats are outside the law the moment they leave their owner's premises, or as soon as they have passed beyond a certain radius from a building. In certain German cities cats are licensed also, but have no rights when they have passed certain limits. Herr Friedrich Schwabe, head of the von Berlepsch School of Bird Protection at Seebach,

---

[1] The Cat, Past and Present, translated from the French of M. Champfleury, with notes by Mrs. Cashel Hoey, 1885, pp. 65, 66.

writes as follows to Mr. William P. Wharton of Groton (translated from the German): —

The law for killing roaming cats varies according to whether it is carried out by those empowered to do so or by owners without authorization. The former may, without further ceremony, shoot any cat, whether roaming wild or not, which they find on their beat, no matter whether the owner is known to them or not. But they [the shooters] must keep a certain distance away from any inhabited building, this distance varying in different States [usually it amounts to 200 metres]. In most domains, those having the legal right to shoot may even demand a fee from the owner of the cat, which fee the owner must pay. The owner of a garden or park who has suffered damage on account of birdcatching cats need only refer to paragraph 228 of our code of civil law if he wishes to legally justify the killing of cats. "After this any one who harms or destroys a foreign object in order to ward off threatened danger from himself or from some other person does not commit an illegal act, provided the harm or destruction is necessary for warding off the danger, and provided the damage is not out of proportion to the danger." Applied to the cat question that means: The owner of a garden in which birds brood may kill cats appearing there if he is able to prove that these cats prey upon the birds and their broods. To be sure, judicial decisions unfavorable to owners of gardens, these owners having killed cats, are not lacking. But in these cases there were culpable accessory circumstances, such as the use of firearms without a permit, or inadmissible nearness to inhabited buildings.

Our laws are unquestionably inadequate, and for that reason the government and the representatives of the people will very soon be obliged to take new measures for the protection of birds.

The experiment of taxing cats has also been tried in order to reduce their number, but this measure has been taken only by towns, and the result cannot yet be seen.

An important point of view is given, in any event, by the fact that the domestic cat — with you in America as well as here with us — cannot be considered and esteemed a native animal belonging to the lineal fauna, but that it is an imported stranger which one can justly return to the house of its owner. There is no reason why the privilege of roaming about freely, denied other domestic animals, should be given to the cat.

According to Dr. Clifton F. Hodge this is practically the solution of the problem reached by Baron von Berlepsch in Germany, and there cities provide traps which are continually kept baited and set for stray cats. According to this writer Hamburg has 300 such traps that during the three years previous to the publication of his book had rid the city of 6,226 cats. He mentions Berlin, Hamburg, Elberfeld, Barmen, Frankfort, Lüneburg, Nuremberg, Pirna, Oels, Breslau, etc., as making official provision for the destruction of cats, and states that in Munster there has existed for some years an "Anti-Cat Society" which has already destroyed several thousand of these "beasts of prey."

In Europe the cat owner seems to have been defeated in the

higher courts. In America the owners of domesticated animals have their rights defined by law, but the status of the cat seems to have been determined largely by the opinion of the presiding justice, who may regard it as domesticated or as a wild animal.

The following is an extract from a newspaper report of a portion of the decision of Judge Utley of Worcester in a case where Dr. Dellinger was arraigned for injuring and destroying cats that were molesting birds that he was engaged to care for: —

> A cat is a wild animal. There is no wilder animal in Christendom. It is an animal that can't be controlled and you can't tell what it will do when it gets out of its owner's sight. A man on his own property has a right to protect it, and when wild animals encroach on it, he is justified in getting rid of them. I find on the evidence presented in this case that the defendant was justified in doing what he did. I don't mean, however, to assert that a man has the right to throw stones promiscuously any place. The defendant is discharged. (Judge Samuel Utley, Criminal Session of the Central District Court, *in re* Thomas Butler *v.* Dr. Oris P. Dellinger. "Worcester Evening Post," Sept. 27, 1905.)

There is a later decision in Maine which is favorable to the cat, but the circumstances were reversed, as the owner of the cat was the defendant.

The following appears in the "Rural New Yorker:" —

> A man in Maine owned a valuable fox terrier dog which went upon a neighbor's property and chased a cat. While it was doing so the owner of the cat shot the dog and killed it. The dog's owner sued the neighbor for damages, and won a verdict on the ground that the cat is not a domestic animal and therefore not entitled to legal protection. . . . The cat owner was not satisfied and appealed the case, his lawyer making a long argument to show that the cat is even more a domestic animal than a dog. He succeeded, and the court reversed the lower verdict, which means that the cat owner was justified in protecting his property. He apparently had as much right to kill a dog which chased his cat as he would have in the case of dogs found worrying sheep.

It will be noted that in both the above cases the owner of the property or his agent were sustained. A man killing another's cat or dog on his own property may have some legal rights that he might not claim in killing it on the owner's property. Malicious killing probably would be unlawful also, as it might come under the head of malicious mischief, and cruelty must be avoided. Dr. Henry Hall of Binghamton, N. Y., was convicted June 8, 1912, before Judge Albert Hotchkiss of the City Court of Binghamton, apparently not for killing a cat, but for failing to kill it and leaving it to suffer. The doctor shot, with a rifle, a cat that was attempting to kill a bird at his drinking fountain, and left it for dead, without taking means to determine whether it was dead or alive. The cat returned to consciousness with its jaw broken, and crawled away. The doctor was fined $25, appealed the case to the County Court of Broome County before Judge Parsons, and there the conviction was sustained Dec. 27, 1912. This seems to have been a conviction for cruelty to animals. Had the cat been shot dead the plaintiff would have had no case. Appolinary Kane of Binghamton was sentenced by Judge Hotchkiss in July, 1915, to thirty days in jail for shooting a cat which he claimed had been killing his chickens. The shot mutilated the cat, and Mr. Kane then went into the house and left the cat to die in agony. It behooves those who shoot cats to beware of bungling and unnecessary cruelty, and to finish the task if they begin it. But there seems to be no law to prevent the humane killing of stray cats anywhere, unless one breaks laws against shooting within city limits, within a certain distance of a dwelling, on the public highway or on public lands; provided also that the trespass laws are not broken in the act. Those who intend to poison or trap cats in Massachusetts should observe the provisions of chapter 626 of the Acts of 1913, which reads as follows: —

SECTION 1. Whoever shall place or distribute poison in any form whatsoever, for the purpose of killing any animal, or shall construct, erect, set, repair or tend any wire snare for the purpose of catching or killing any animal, shall be punished by a fine of not exceeding one hundred dollars: *provided*, that nothing in this section shall be construed to prohibit any person from placing in or near his house, barns or fields, poison intended to destroy rats, woodchucks or other pests of a like nature or insects of any kind.

SECTION 2. Any person who shall set, place, maintain or tend a steel trap with a spread of more than six inches or a steel trap with teeth jaws, or a "stop-thief" or choke trap with an opening of more than six inches shall be punished by a fine of not exceeding one hundred dollars.

SECTION 3. Any person who shall set, maintain, or tend a steel trap on enclosed land of another without the consent in writing of the owner thereof, and any person who shall fail to visit at least once in twenty-four hours, a trap set or maintained by him shall be punished by a fine of not exceeding twenty dollars.

Section 70, chapter 212, Revised Laws (1902), provides a penalty for cruelly abandoning any domestic animal. Only a few convictions for deserting cats have been secured under this law for the reason that it often is hard to prove which has been abandoned, cat or owner.

RECAPITULATION AND CONCLUSION.

The cat was domesticated within historic times, but did not appear as an inmate of the home in western Europe until about 900 A.D. Civilized man managed very well without it for centuries. Puss appears to have been domesticated first in Egypt about 1200 to 1600 B.C. by the taming of certain wild African species.

The household pets of to-day are believed to have descended from African, Asiatic and European species.

The cat is far more widely kept and distributed than any other domestic animal, and is under less control and restraint than any other. It usually has a greater affection for places than for persons, and tends to return to its home when its owner moves away. Also, it readily abandons its owner, and, often abandoned by him, returns to the wild. Incalculable numbers of wild or stray house cats now roam the woods and fields of New England. These wild cats attract others from their homes.

Many, remaining with the owners, are fed insufficiently or not at all, and having to rely on their own efforts for food, emulate those that have run wild. Many pet cats are allowed to roam the country at night. People keep too many cats, and as the population increases the number of cats increases accordingly.

The cat, an introduced animal, is not needed here outside of buildings. It has disturbed the biological balance and has become a destructive force among native birds and mammals. It is a member of one of the most bloodthirsty and carnivorous families of the mammalia, and makes terrific inroads on weaker creatures. It is particularly destructive to certain insect-eating forms of life, such as birds, moles, shrews, toads, etc. Every year the cats of New England undoubtedly destroy millions of birds and other useful creatures, therefore indirectly aiding the increase of insects which destroy crops and trees. Such insects possibly cost the people of Massachusetts from seven and one-half million to nine million dollars annually. The cat protects them, thus increasing the cost of living to every citizen. The good that cats accomplish in the destruction of field mice, woods mice and insects is of little consequence beside the ravages that they inflict among insectivorous birds and other insect-eating and mouse-eating creatures.

Cats, selected for their rat-killing propensities, are useful if kept in their proper place in and around buildings, but the species is so destructive to game and to valuable wild life that it should not be allowed to roam, particularly in the country.

City cats should not be taken to the country in the summer and there permitted to run at large, to prey on birds and game, nor should they be abandoned and left to their own devices at the close of the season. This is both cruel and unlawful.

Many people do not keep cats. Rats and mice are disposed of by ratproofing buildings and food receptacles and using traps. (See Economic Biology Bulletin No. 1, "Rats and Rat Riddance," published by the Massachusetts State Board of Agriculture.) The utility of the cat in destroying rats and mice has been both overrated and understated. The testimony of cat lovers and cat owners, taken during a canvass in several counties of Massachusetts, seems to indicate that only about one-third of the cats kept in the country towns are known to catch rats, and that only about one-fifth of them are efficient ratters. The number of mousers is larger, but mice may be readily disposed of by traps. It is probable that one-fifth of the cats kept in the country, properly selected and restrained, would accomplish as much in killing rats and mice as do those now kept, and possibly the requisite number might be still further reduced by careful selection and breeding.

Apparently the cat has few legal rights. In most countries the law seems to regard it as a predatory animal which any person may destroy when found doing damage on his premises. In Massachusetts and some other States the laws protect it from cruelty and abuse. People killing cats should observe all laws or ordinances in regard to trespassing, cruelty, shooting, trapping or poisoning. A cat apparently has some rights on the property of its owner that are denied it when on the property of others.

There are laws to protect insectivorous birds against gunners, snarers and trappers. Birds of prey and wild predatory animals are proscribed by law, and bounties are offered on the heads of some. Many States offer bounties for native wild cats, but there is no law to check the ravages of the wild house cat, — a far more numerous animal. A man may be fined $10 for killing a songbird, but he may keep any number of cats and may train them to kill many birds weekly. Hardly a hand is raised to stay the destruction of valuable wild life by hundreds of thousands of vagabond or wild house cats. Hunters and trappers have little incentive to kill them as the fur is of small value. Legislation is needed to check this evil.

It is undeniable that cats may carry such infections as smallpox and scarlet fever, but the subject requires careful investigation before exact statements can be made. The evidence thus far offered is inconclusive. Cats undoubtedly disseminate ringworm, and rabies in the cat is more dangerous to man than in the dog, but rarer. In some cases serious infections appear to have been transmitted by the bites or scratches of cats, but here again the evidence of direct infection is not conclusive, as any wound may become infected after infliction.

The evils connected with the unrestrained liberty of the cat can be abated only by reducing the number of cats to a minimum, limiting breeding, destroying superfluous kittens at birth, restraining or confining cats kept as pets and as ratters (particularly at night and during the breeding season of the birds), quarantining cats in cases of infectious diseases, and destroying all stray and feral cats, wherever they may be found.

When it becomes necessary to allow barn cats free range, that they may destroy rats outside of buildings during the summer months, they should be supplied with water and well and regularly fed with meat and other animal foods. Probably in most cases they will then be less likely to roam the fields and more inclined to lie in wait for rats and mice than if not well fed.

In dealing with the cat from an economic point of view we need raise no question of the rights of the animal. Man has won his way upward through the great struggle by his own powers of mind out of prehistoric darkness to the place of command. He now controls the destinies of his fellow creatures. He may concede them certain rights only if such concession does not interfere with the best interests of all.

Animals were domesticated because of their utility to man in his struggle upward from savagery. The sympathy which he feels for his helpers and pets, praiseworthy and important as it is, is a secondary consideration. The claims of the cat to a place in our domestic life rest primarily on the fact that it is supposed to do for us, with little conscious effort on our part, the onerous, petty and disagreeable task of destroying small rodents which for centuries have elected to fasten themselves as parasites on civilization. Insomuch as the creature fails in this, in so far as it destroys other more useful or nobler forms of life, in such measure it becomes an evil and a pest. It will become an influence for good or ill according as we mould it, restrain it and limit its activities. It is our duty to check, with a firm hand, its undue increase in domestication, and to eliminate the vagrant or feral cat as we would a wolf.

LIST OF THOSE WHO CONTRIBUTED INFORMATION.

Adams, Emily B., Springfield.
Adams, William C., Boston.
Affleck, G. B., Springfield.
Aiken, Mary A., Norwich, Conn.
Allen, Willis Boyd, Boston.
Ambrose, David A., Newton.
Ames, J. S., Gardner.
Anthony, B. W., Adrian, Mich.
Aspinwall, Thomas, Brookline.
Atherton, Edward H., Roxbury.
Atkinson, H. R., Brookline.
Averill, Florence M., North Andover.
Avery, Frederick L., Ayer.
Ayres, Mary A., Medford.
Babson, Caroline W., Pigeon Cove.
Bagnall, F. A., Adams.
Bailey, Dr. Bernard A., Wiscasset, Me.
Bailey, S. Waldo, West Newbury.
Baker, Lorenzo D., Jr., Boston.
Ballard, Geneva S., Millington.
Bancroft, Alice W., Brookline.
Bancroft, W. F., Washington, D. C.
Barber, John W., Newton.
Barlow, Richard H., Methuen.
Barnard, Rev. Margaret B., Rowe.
Barnes, Dwight F., Marshfield.
Bartlett, Herbert W., Plymouth.
Bascom, E. A., Georgetown.
Bassett, Thomas J., Leominster.
Bates, F. A., South Braintree.
Battelle, Judson S., Dover.
Beals, Ella M., Marblehead.
Bemis, Benjamin F., Gleasondale.
Bemis, James E., Framingham.
Bent, C. L., Gardner.
Bishop, Dr. Louis B., New Haven, Conn.
Blair, Wesley W., Newtonville.
Blake, B. S., Weston.
Blanchard, William, Tyngsborough.
Boardman, Mrs. H. C., New Bedford.
Bonney, Mrs. A. H., West Hanover.
Bowdish, B. S., Demarest, N. J.
Bowen, A. M., Springfield.
Boyd, Harriet T., Dedham.
Brastow, Amelia M., Wrentham.
Brewer, W. S., Hingham.
Brewster, William, Cambridge.
Bridge, Mrs. Edmund, West Medford.
Briggs, Oliver L., Boston.
Brigham, Margaret, North Grafton.
Brockway, Arthur W., Hadlyme, Conn.
Brooks, S., Boston.
Brooks, Dr. William P., Amherst.
Brown, Annie H., Stoneham.
Brown, C. Emerson, Boston.
Brown, Frank A., Beverly.
Brown, Mrs. Henry T., Lancaster.
Browning, Mrs. Julia F. A., Rowe.
Browning, Wm. H., New York City.
Bruce, C. O., Mt. Hermon.
Bruen, Frank, Bristol, Conn.
Brundage, A. B., Danbury, Conn.
Bryant, H. C., Berkeley, Cal.
Buck, Henry R., Hartford, Conn.
Buckley, Emma, Worcester.
Buffington, Samuel L., Swansea.
Bugbee, Edgar L., Fitchburg.
Burdick, Mabel G., Stapleton, N. Y.
Burgess, Mrs. M. E., Pittsfield.
Burney, Thomas L., Lynn.
Burnham, John B., New York City.
Burns, Frank L., Berwyn, Pa.
Burt, Mrs. J. M., East Longmeadow.
Butler, Mrs. Florence L., East Charlemont.
Cady, Mrs. J. H., Providence, R. I.
Cardee, Jos. H., Bolton.
Carne, Mrs. Thomas, Adams.
Carney, Edward B., Lowell.
Carter, H. S., New Britain, Conn.
Case, Clifford M., Hartford, Conn.
Chapin, Myra F., Granby.
Cheesman, William H., Washington, D. C.
Cheney, Louis R., Hartford, Conn.
Cheney, Rev. R. F., Southborough.
Child, Rev. Dudley R., Pepperell.
Chipman, Grace E., Sandwich.
Church, Elliott B., Newton.
Colburn, David M., Fitchburg.
Cole, Edwin M., Cohasset.
Coney, Kate E., West Roxbury.
Cook, John A., Gloucester.
Coonan, Thomas J., Jr., Worcester.
Corliss, Wm. D., Gloucester.
Couch, Mrs. Franklin, Dalton.
Cowing, D. T., Hadley.
Crampton, John M., Hartford, Conn.
Crandall, Lee S., New York City.
Crockett, Edith B., Brandon, Vt.
Crosby, M. S., Rhinebeck, N. Y.
Currier, Freeman B., Newburyport.
Curtis, Albert E., Ballardvale.
Cushman, E. Wesley, Scituate.
Davidson, Charles S., South Williamstown.
Davis, George, Cambridge.
Day, Chester S., West Roxbury.
Day, F. B., Stoneham.
Day, William, Vineyard Haven.
Deane, Daniel W., Fairhaven.
Decker, Harold K., West New Brighton, N. Y.
DeCosti, Edward, Dedham.
Dewey, Dr. Chas. A., Rochester, N. Y.
Dixon, Francis E., Eliot, Me.
Dixon, Frederick J., Hackensack, N. J.
Donaldson, Geo. C., Hamilton.
Donlon, Henry J., Fitchburg.
Dorman, Albert X., Worcester.
Drew, Miss Evie W., Hanson.
Dumbell, Rev. Howard M., Delhi, N. Y.
Dutcher, William, Plainfield, N. J.
Dyke, Arthur C., Bridgewater.

Eames, Agnes C., Wilmington.
Eastman, Alfred C., Westwood.
Eastman, George F., Granby.
Eastman, Harry D., Sherborn.
Eaton, Charles E., Orange, N. J.
Eddy, Newell A., Bay City, Mich.
Eldredge, Hattie D., East Falmouth.
Elliot, Mrs. J. W., Boston.
Ellis, Cyril F., Fitchburg.
Emery, Georgia H., Newton.
Ensign, Chas. S., Newton.
Fairbanks, Mrs. Edward T., St. Johnsbury, Vt.
Fales, Wyman E., West Somerville.
Fanning, Dr. W. G., Danvers.
Farley, John A., Plymouth.
Farrar, Hilda, Rochester, N. Y.
Farwell, Leon C., Fitchburg.
Faunce, Sewall R., Dorchester.
Fearing, Mary P., Boston.
Felton, T. P., West Berlin.
Field, Mrs. Charles M., Shrewsbury.
Field, Dr. George W., Sharon.
Fisher, Dr. A. K., Washington, D. C.
Fletcher, Emily F., Westford.
Fottler, John, Dorchester.
Fowler, Mrs. E. S., Danvers.
Frost, Cornelia, Boston.
Fuller, Annie A., Kingston.
Fuller, William, Auburndale.
Gaylord, E. E., Beverly.
Gerard, Mrs. F. W., South Norwalk, Conn.
Goddard, Mrs. H. L., Shrewsbury.
Gold, C. L., West Cornwall, Conn.
Goldsmith, Gertrude B., Manchester.
Goldthwait, Mrs. Chas. S., Peabody.
Goodhue, Charles F., Penacook, N. H.
Goodwin, Frederick W., East Boston.
Goodwin, James, Hartford, Conn.
Gordon, J. Wilson, Yonkers, N. Y.
Gorst, Charles Crawford, Boston.
Gould, Alfred M., Malden.
Grant, Carl E., Gloucester.
Graves, S. P., Walpole.
Gray, George M., Woods Hole.
Greene, Caroline S., North Cambridge.
Greenlaw, Henrietta, Dedham.
Gregory, Herbert, Leominster.
Grennan, Miss G. B., Woodberry Forest, Va.
Grout, A. J., New Dorp, N. Y.
Hager, George W., Marlborough.
Hale, Richard W., Dover.
Handy, Mrs. Louise H., Marion.
Hanson, Ray F., Fitchburg.
Hardin, Alfred B., Foxborough.
Harriman, Rev. Frederick W., Windsor, Conn.
Hartman, Edward T., Allston.
Hastings, George H., Fitchburg.
Haynes, Elizabeth C., Brookline.
Hayward, Anna R., Melrose.
Headley, P. C., Jr., Fairhaven.
Hemenway, Mrs. Augustus, Boston.
Henderson, Alexander, Brookline.

Henderson, Jessica L. C., Wayland.
Henderson, Walter P., Dover.
Herrick, Harold, Lawrence, N. Y.
Higgins, Myrta M., Framingham.
Hildreth, Mrs. Fannie B., Northborough.
Hittinger, Jacob, Belmont.
Hoar, Samuel, Concord.
Hobbs, Lewis F., West Medford.
Holden, E. F., Melrose.
Holmes, George B., Kingston.
Honywill, A. W., Jr., Wilkinsburg, Pa.
Hornaday, Dr. William T., New York City.
Hornbrooke, Mrs. Francis B., Newton.
Howard, Anson O., East Northfield.
Howard, Emma L., South Easton.
Howard, J. S., Provincetown.
Howe, L. H., Newton.
Howe, R. Heber, Jr., Concord.
Howes, Helen E., Boston.
Hoxsie, George E., Canonchet, R. I.
Hubbard, George F., Fitchburg.
Hubbard, Marian E., Wellesley College.
Huntington, R. W., Jr., Hartford, Conn.
Hutchins, Charles L., Concord.
Hutchinson, Calvin B., Whitman.
Hylan, Rev. Albert E., Medfield.
Jacobs, Eliza C., West Roxbury.
Jefts, Arthur W., Worcester.
Jenks, Caroline E., Bedford.
Jenks, Charles W., Bedford.
Jensen, Christian E., Fitchburg.
Jensen, J. K., Westwood.
Jewett, Elizabeth, Yarmouthport.
Johnson, Byron B., Waltham.
Johnson, E. Colfax, Shutesbury.
Jones, Abby B., Kingston.
Jones, Jonathan H., Waquoit.
Jones, Dr. L. C., Falmouth.
Jones, William F., Norway, Me.
Jones, William H., Nantucket.
Jones, William Preble, Somerville.
Kane, Charles M., Spencer.
Kane, John F., Fitchburg.
Kemp, Parker J., Pepperell.
Keniston, Allan, Edgartown.
Kennedy, Mrs. Augusta M., Whitman.
Kenney, James W., Somerville.
Keyes, Mrs. Prescott, Concord.
King, Henry B., Augusta, Ga.
Kinney, Henry E., Worcester.
Kittredge, Harold W., Leominster.
Klinger, Bertha H., Hartford, Conn.
Knowlton, S. Everett, Wenham.
Ladd, Mrs. Geo. S., Sturbridge.
Lakeman, Sarah E., Ipswich.
Lantz, Prof. David E., Washington, D. C.
Larkin, Walter A., Andover.
Latham, Charles R., Suffield, Conn.
Laurent, Philip, Philadelphia, Pa.
Learned, A. K., Gardner.
Leighton, Helen, Fall River.
Leland, Ernest M., Fitchburg.
Leonard, Eliza B., Greenfield.
Leonard, William H., East Foxborough.
Levey, Mrs. William M., West Hartford, Conn.

Lewis, Hershel W., New Ipswich, N. H.
Lewis, J. B., Reading.
Linton, Morris, Moorestown, N. J.
Livermore, Perkins R., Marshfield Hills.
Lloyd, Mrs. A. W., Wakefield.
Locke, A., Tottenville, N. Y.
Loveland, Clifton W., Providence, R. I.
Luman, John F., Thorndike.
Lundigen, Ralph J., Leominster.
Lusk, Mrs. Louise G., Unionville, Conn.
Lyman, A. M., Montague.
Macy, William F., West Medford.
Malley, John F., Fitchburg.
Mann, James R., Arlington Heights.
Manning, Warner H., Boston.
Mansfield, Helen, Gloucester.
Marsh, Dr. Franklin F., Wareham.
Marshall, Mrs. E. O., New Salem.
Marston, Howard, Barnstable.
Martin, R. O., Lenoxdale.
Mason, Vinton W., Cambridge.
Matthews, C. F., Shutesbury.
Maxwell, Mrs. Paul S., Pepperell.
May, Dr. John B., Waban.
May, John F., Fitchburg.
Maynard, Mrs. Amy B., Northborough.
Maynard, C. J., West Newton.
McAndrews, Walter F., Fitchburg.
McCaffrey, Joseph, Clinton.
McCue, Hugh, East Milton.
McIntosh, Mrs. Frederick, Nahant.
McKittrick, Frank G. W., Tyngsborough.
McLean, J. B., Simsbury, Conn.
McRae, Mabel, Boylston.
Meech, H. P., West Hartford, Conn.
Meier, W. H. D., Framingham.
Mellus, J. T., Wellesley.
Meredith, Mrs. Albert A. H., Milton.
Merrill, Albert R., Boston.
Meyer, Heloise, Lenox.
Miles, Mrs. Henry A., Hingham.
Miller, Charles A., Walpole.
Minot, William, Boston.
Mirick, George D., Stoneham.
Monahan, Peter P., Westfield.
Moran, Charles, Clinton.
Moran, John F., Clinton.
Morris, Charles, New Haven, Conn.
Morris, George E., Waltham.
Morris, Mrs. James F., Providence, R. I.
Morse, C. Harry, Belmont.
Morse, Eliza A., Worcester.
Morse, Frank E., Auburndale.
Moseley, Charles W., Newburyport.
Mosher, F. H., Melrose.
Moulton, Rev. J. Sidney, Stow.
Munns, Dr. C. O., Oxford, O.
Murphy, Robert Cushman, Brooklyn, N. Y.
Newton, Dr. Carrie E., Brewer, Me.
Nichols, Mary W., Danvers.
Norcross, Otis W., Baldwinville.
Northey, William E., Topsfield.
Norton, Arthur H., Portland, Me.
Nutt, N. A., South Ashburnham.
Olney, William B., Seekonk.
Otis, Herman, Fitchburg.
O'Toole, John, Clinton.
Otterson, A. W., Hall, N. Y.
Packard, Anna W., Hudson.
Packard, Winthrop, Canton.
Parker, Augustin H., Charles River Village.
Parker, Harold, Lancaster.
Parker, Herbert, Lancaster.
Paxon, Mrs. A. M., Lowell.
Peabody, Charles J., Topsfield.
Pease, Mrs. Cora E., Malden.
Pease, E. Linn, Thompsonville, Conn.
Pease, Harriet R., Greenfield.
Peaslee, Frank J., Lynn.
Perron, Homer E., Worcester.
Perry, Dr. Henry J., Boston.
Phillips, Geo. G., Green, R. I.
Phypers, Mrs. G. W., South Euclid, O.
Pierce, George Willis, Jamaica Plain.
Pilsbury, Frank O., Walpole.
Piper, George W., Andover.
Pitman, Harold A., Boston.
Poole, J. Edward, Lynn.
Pope, Alexander, Brookline.
Porter, Juliet, Worcester.
Powell, Edwin C., Springfield.
Powell, Mrs. S. W., Great Barrington.
Powers, L. Moore, Gloucester.
Pratt, Edward H., North Adams.
Pratt, Nathan W., North Middleborough.
Prescott, C. W., Concord.
Puffer, Loring W., Brockton.
Pulley, James M., Melrose.
Pulsifer, William H., Pittsfield.
Quinby, Bertha W., Saco, Me.
Rawson, Charles I., Oxford.
Redfield, Julia W., Pittsfield.
Rice, George H., St. Augustine, Fla.
Rich, Mrs. Snow, Boston.
Richards, Harriet E., Brookline.
Richardson, Clarence E., Attleboro.
Richardson, Guy, Dorchester.
Richardson, John K., Wellesley Hills.
Robbins, Mrs. Reginald C., Hamilton.
Robbins, Reginald C., Hamilton.
Robbins, Samuel D., Belmont.
Robertson, Sylvester P., Plainfield.
Robinson, John, Salem.
Robinson, William A., Tisbury.
Rogers, Howard P., Framingham.
Rogers, J. Riley, Byfield.
Ross, Helen W., Ipswich.
Rountree, H. H., Randolph.
Ruberg, Lyman E., Greenfield.
Rugg, Harold G., Hanover, N. H.
Ruggles, Deane F., Plainfield, N. H.
Saltonstall, John L., Beverly.
Saunders, Mary T., Salem.
Sawyer, Miss M. E., Walpole.
Schaff, Morris, Southborough.
Seabury, Joseph S., Wayland.
Seton, Ernest Thompson, Greenwich, Conn.
Shattuck, Clara M., Pepperell.
Shaw, C. F., Abington.

Shaw, Dr. J. Holbrook, Plymouth.
Shedd, Albert Edward, Sharon.
Sherman, Althea R., National, Ia.
Shumway, Franklin P., Melrose.
Simms, Mrs. Herman E., Haverhill.
Sims, William Fisher, Saugus.
Sinclair, J. A., New Hampton, N. H.
Sitgreaves, Miss M. J., Chestnut Hill.
Slade, Elisha, Somerset.
Slocum, William H., Jamaica Plain.
Small, E. L., North Truro.
Smith, Curtis Nye, Newton.
Smith, W. A., Wilmington.
Smith, Wilbur F., South Norwalk, Conn.
Soule, Caroline G., Brookline.
Stanley, Mrs. Mary R., North Attleborough.
Starbuck, Margaret C., Jamaica Plain.
Starks, Charles E., Winter Hill.
Stevens, F. E., Somerville.
Stevens, Mabel E., St. Johnsbury, Vt.
Stevens, Mabel T., Wollaston.
Stevens, Dr. R. B., Roslindale.
Stiles, Jas. T., Gardner.
St. John, Edward P., Hartford, Conn.
St. John, George C., Wallingford, Conn.
Stockwell, Wallace E., Fitchburg.
Stone, Clayton E., Lunenburg.
Stone, Mrs. F. H., South Dartmouth.
Streeter, Mrs. A. W., Winchendon.
Sturgis, S. W., Groton.
Tenney, Sanborn, Williamstown.
Thayer, Abbott H., Monadnock, N. H.
Thayer, Herbert E., Springfield.
Thompson, Ella W., Woburn.
Till, William, Magnolia.
Tilton, Louis O., Waban.
Tinkham, Horace W., Touisset.
Torrey, Harry A., East Sandwich.
Townsend, Rev. Manley B., Nashua, N. H.
Tucker, William F., Worcester.
Tuttle, Paul G., Fitchburg.
Van Huyck, J. M., Lee.
Vardon, Ross, Greenwood.
Wade, Mrs. Martha, Mansfield.
Wait, Francis A., Medford.
Waite, J. W., South Hadley.
Waite, Margaret L., Cambridge.
Waldo, Chas. Sidney, Jamaica Plain.
Walker, Helen, Milton.
Ware, Lyman E., Norfolk.
Warner, R. L., Concord.
Warren, William A., Lunenburg.
Watson, Frank E., Haverhill.
Weeks, W. B., Beverly.
Wentworth, Nathaniel, Hudson, N. H.
Wharton, William P., Groton.
Wheat, Mrs. Mary A., Dorchester.
Whitcomb, Mrs. Henry F., Amherst.
White, Grace C., West Brookfield.
White, Dr. James C., Boston.
White, Mary A., Heath.
Whiting, Adrian P., Plymouth.
Whiting, Willard C., Cambridge.
Whitmore, Martha W., Plymouth.
Whittaker, Albert E., Fitchburg.
Wilder, Dr. Burt G., Brookline.
Wilder, Grace E., East Lynn.
Willard, Helen, Brookline.
Williams, Dr. Edward R., Cambridge.
Williams, M. P., Wellesley.
Williams, Mrs. Rob't W., Medfield.
Wilson, Francis J., Fitchburg.
Witherbee, Anne F., Marlborough.
Wood, J. Elmer, Beverly.
Woodward, Harry W., Lynn.
Woodward, Dr. L. F., Worcester.
Worthen, Dr. C. F., Weston.
Wright, Mrs. Mabel Osgood, Fairfield, Conn.
Wright, Samuel B., Fitchburg.
Wright, Mrs. Theodore F., Cambridge.
Wyman, Mrs. H. A., Boston.

PLATE XX.

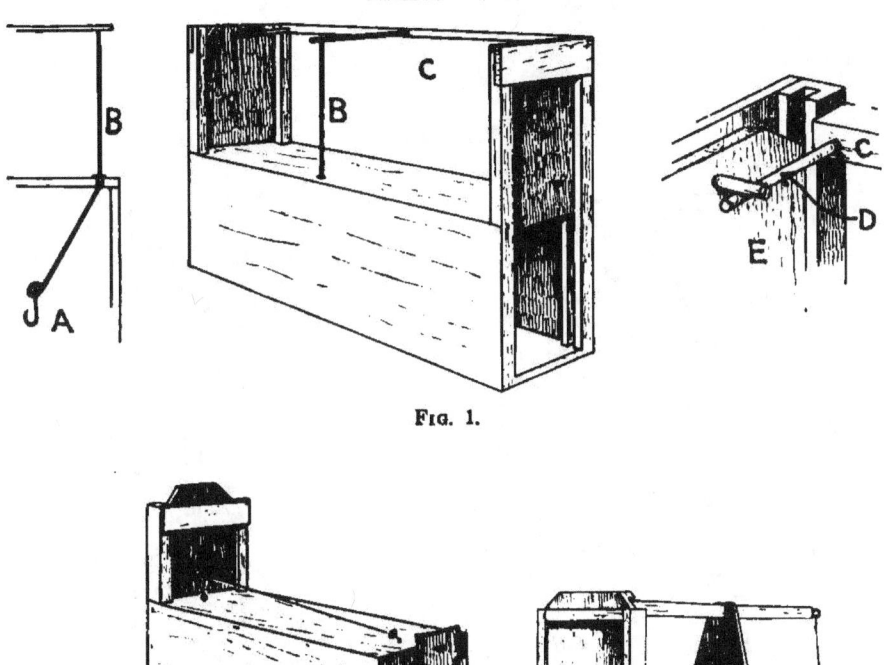

FIG. 1.

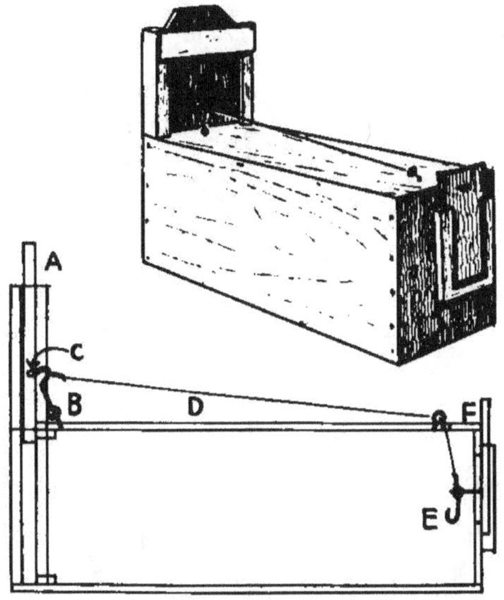

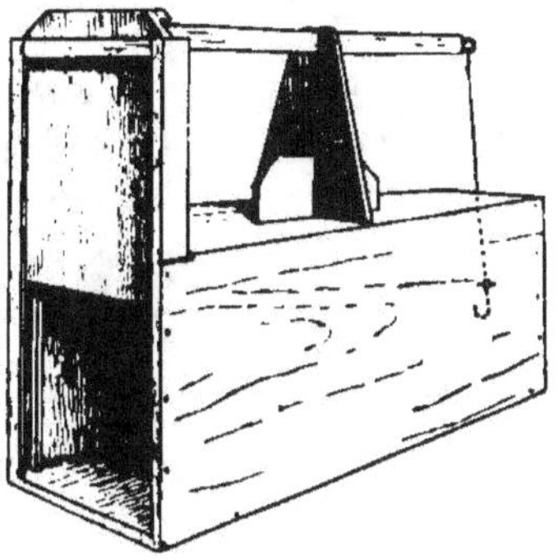

FIG. 2.

FIG. 3.

FIG. 1.—DOUBLE-ENDED TRAP FOR CATS.

Made by Mr. E. E. Edmanson, Chicago.

A. Bait-hook.
B. Trigger-rod of heavy wire.
C. Square rod, loosely pivoted at ends.
D. Rod to support door.
E. Sliding door.

FIG. 2.—SCUDDER CAT TRAP.

Made by Massachusetts Fish and Game Protective Association.

A. Sliding door.
B. Hook supporting door.
C. Hole in door to engage hook.
D. Cord or wire.
E. Bait-hook caught on point of nail.
F. Small door for setting trap and examining contents.

FIG. 3.—DODSON CAT TRAP.

Made by Mr. Joseph H. Dodson, Chicago.

The sliding door is supported by the pivoted lever.
The bait-hook is held lightly on the point of a nail.

PLATE XIX.

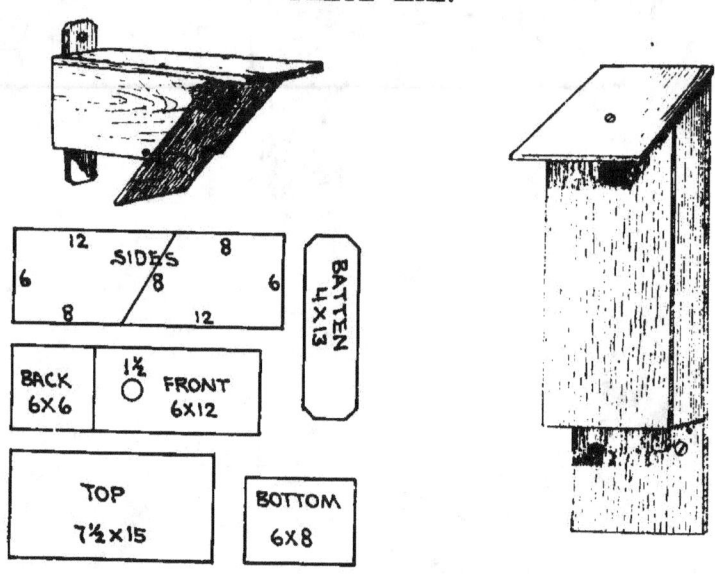

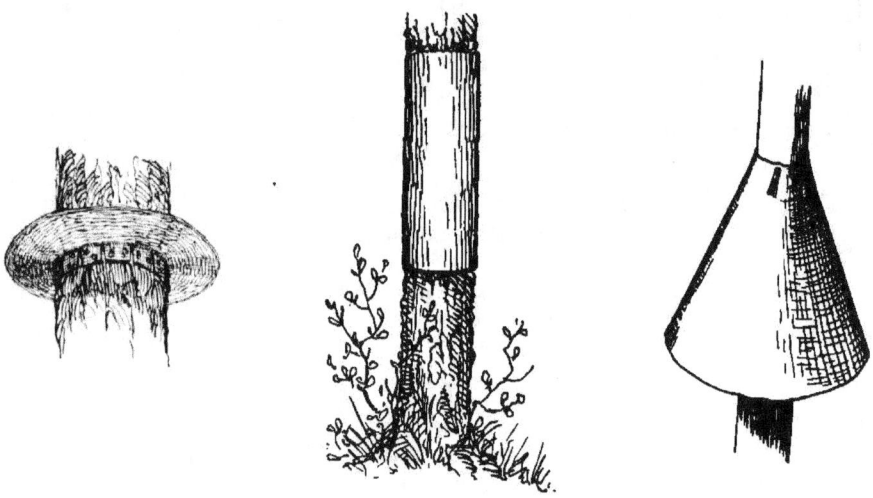

DEVICES TO PROTECT BIRDS' NESTS.
Upper figures show catproof nesting boxes. Lower figures, zinc cat guards for trees or poles.

PLATE XVIII.

FIG. 1. — A CAT WHICH HAS NEVER CAUGHT A BIRD.

This cat, belonging to Dr. Burt G. Wilder, is kept in or caged during the night, fed regularly, and given a good breakfast before his morning liberty. Birds do not interest him. (Photograph by courtesy of Dr. Wilder.)

FIG. 2. — BUSTER, PROUD OF HIS TETHER.

This great cat, owned by Mr. Bardwell Gladwin of Plainville, Conn., is kept tethered to an overhead wire. He has been tied every summer, and seems to consider the collar and leash as a high honor. (Photograph by courtesy of Mrs. Louise G. Lusk.)

PLATE XVII.

DOROTHY PERKINS ROSEBUSH.
Trained on pole to prevent cats from climbing to bird house. (After "Our Dumb Animals.")

PLATE XVI.

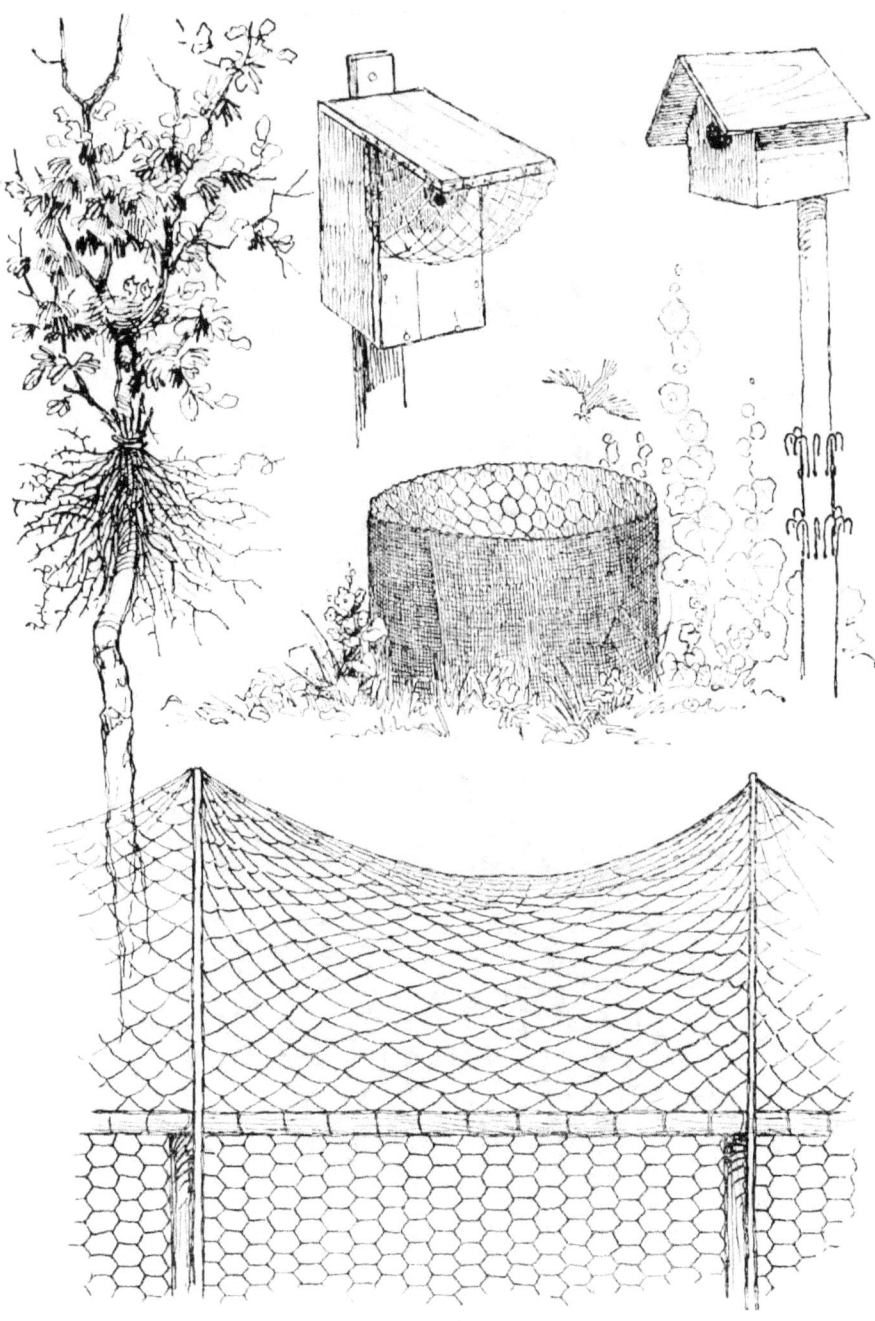

DEVICES FOR PROTECTING BIRDS, THEIR NESTS AND YOUNG.

Upper figures show protectors for birds' nests on tree, pole, and ground. Lower figure, catproof fence topped by a fish net. This is a success. (See pages 88 and 93.)

PLATE XV.

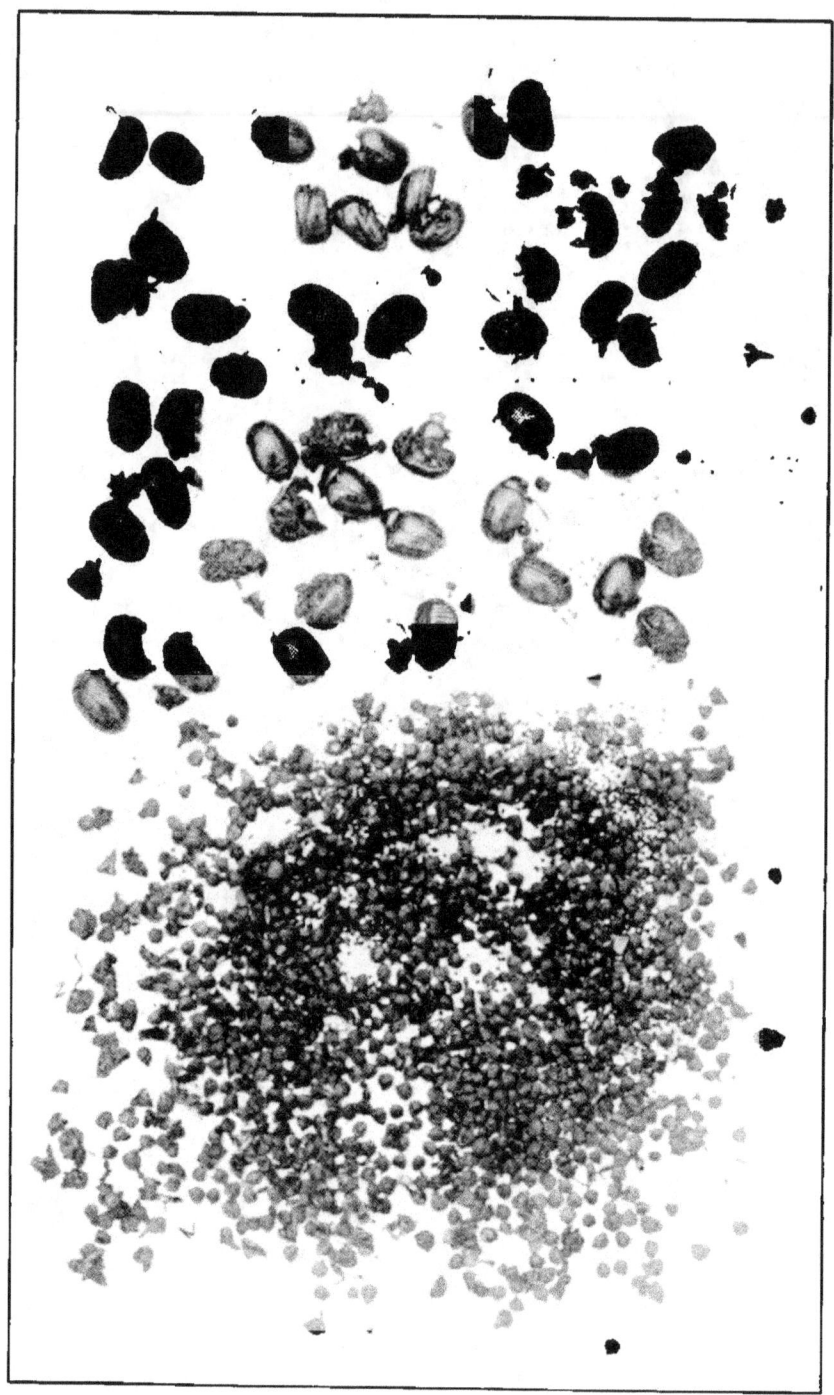

CONTENTS OF A BOBWHITE'S CROP.
Forty-eight potato beetles and about two hundred and fifty weed seeds. This does not include the contents of the stomach. (Original photograph.)

PLATE XIV.

Fig. 1. — The Cat kills the Bluebird on its Nest.
Female bluebird with weevil. Weevils destroy grain, fruit and vegetables. The bluebird is very useful. (Original photograph.)

Fig. 2. — Every Bluebird a Help to the Farmer.
Male bluebird with grasshopper. Many bluebirds are killed yearly by cats. (Original photograph.)

PLATE XIII.

Fig. 1. — A Common Victim of the Cat.

The cat kills the chickadee, one of our most useful birds. Note the caterpillar in its bill.
(Original photograph.)

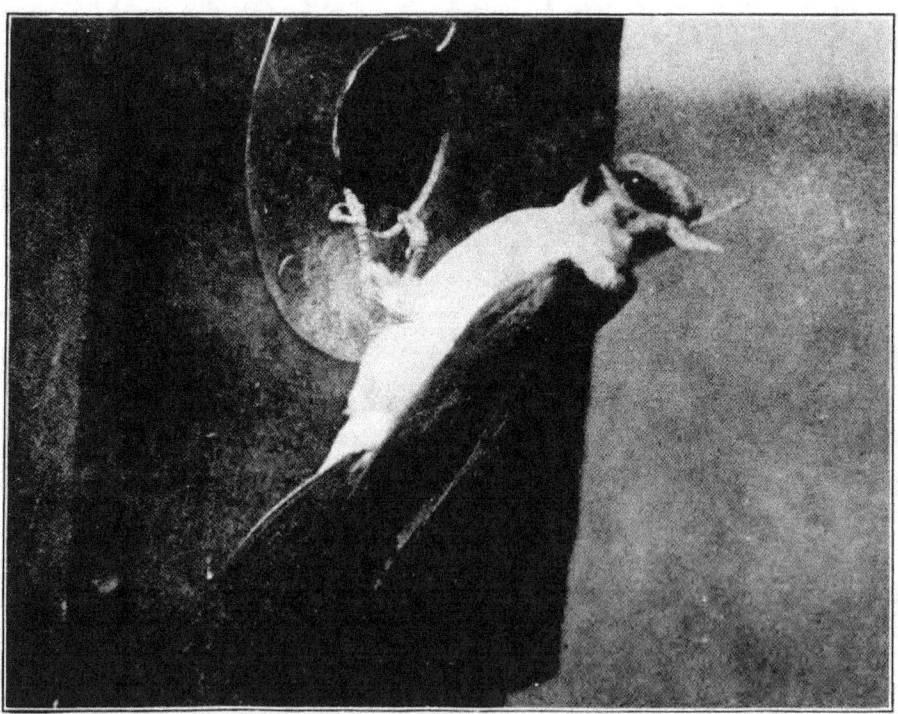

Fig. 2. — Another of the Cat's Victims.

A cat killed nine tree swallows in one day. This bird's throat is packed with insects and the wings of a cutworm moth protrude from its mouth. (Original photograph.)

PLATE XII.

FIG. 1. — AN ILLUSTRATION OF THE INEFFICIENCY OF THE CAT AS A RATCATCHER.

One cat and twenty-four rats, the result of fumigating cabin of steamship. This cat, an exceptionally good ratter, was supposed to have kept the cabin free from rats. In fumigation she was overlooked. (From Public Health Reports, Vol. 29, No. 16.)

FIG. 2. — RAT TRAPS WELL HANDLED BEAT THE CAT.

Twenty-three rats and about a dozen mice trapped in two barns in three days, with 5-cent traps, properly set. The only advantage of the cat as a rat trap is that it is self-setting.

PLATE XI.

A HUNTING CAT AND ITS VICTIM.
This animal feasted on the rabbits and squirrels of the New York Zoölogical Park until it ate only the brains of its victims. (Photograph by courtesy of Dr. Wm. T. Hornaday.)

PLATE X.

THE CAT'S PREY.
Full-grown gray squirrel killed by a cat. (Photograph by courtesy of Dr. Wm. T. Hornaday.)

FIG. 1. — REMAINS OF BIRDS KILLED ON THEIR NESTS BY A WANDERING CAT.
Deputy Fish and Game Commissioner Allan Keniston examining the remains of Wilson's terns at Katama Beach. (Photograph by Mr. Howard H. Cleaves.)

FIG. 2. — THE CAT'S LAIR.
A mass of bird remains on the beach grass at Katama Bay, where a wild house cat had

PLATE VIII.

Fig. 1. — No Cats here.
A nest of Wilson's tern, undisturbed on an island where no cats lived.

Fig. 2. — The Cat's Work. A Wanton Killing.
Remains of a mother tern as found; killed by a cat. Thousands of these birds have been killed on their nests by cats on Muskeget. (Photograph by Mr. Howard H. Cleaves.)

PLATE VII.

FIG. 1. — EXPENSIVE CATS.
Five cats which, it is estimated, cost New York $1,000 by destroying game birds at the State Game Farm. (Photograph by courtesy of Mr. Herbert K. Job. See page 49.)

FIG. 2. — REMAINS OF HEN PHEASANT CAUGHT ON NEST BY A CAT.
This bird was killed and eaten by a cat at 10 P.M., at Wilkinsonville. (From the annual report of the Massachusetts Commission on Fisheries and Game, 1911.)

PLATE VI.

Fig. 1. — Full-grown Ruffed Grouse killed by a Cat on the Snow in East Milton. The cat was driven away and the bird picked up still breathing. (Photograph by Mr. Walt F. McMahon.)

Fig. 2. — Bells on Cats will not save Birds.
A fine, sleek, pet Angora, with six bells on its collar, brought in thirty-two birds during one nesting season and twenty-eight the next. It is shown here killing a young catbird. (Photograph by courtesy of Mr. Neil Morrow Ladd, Greenwich, Conn. See page 93.)

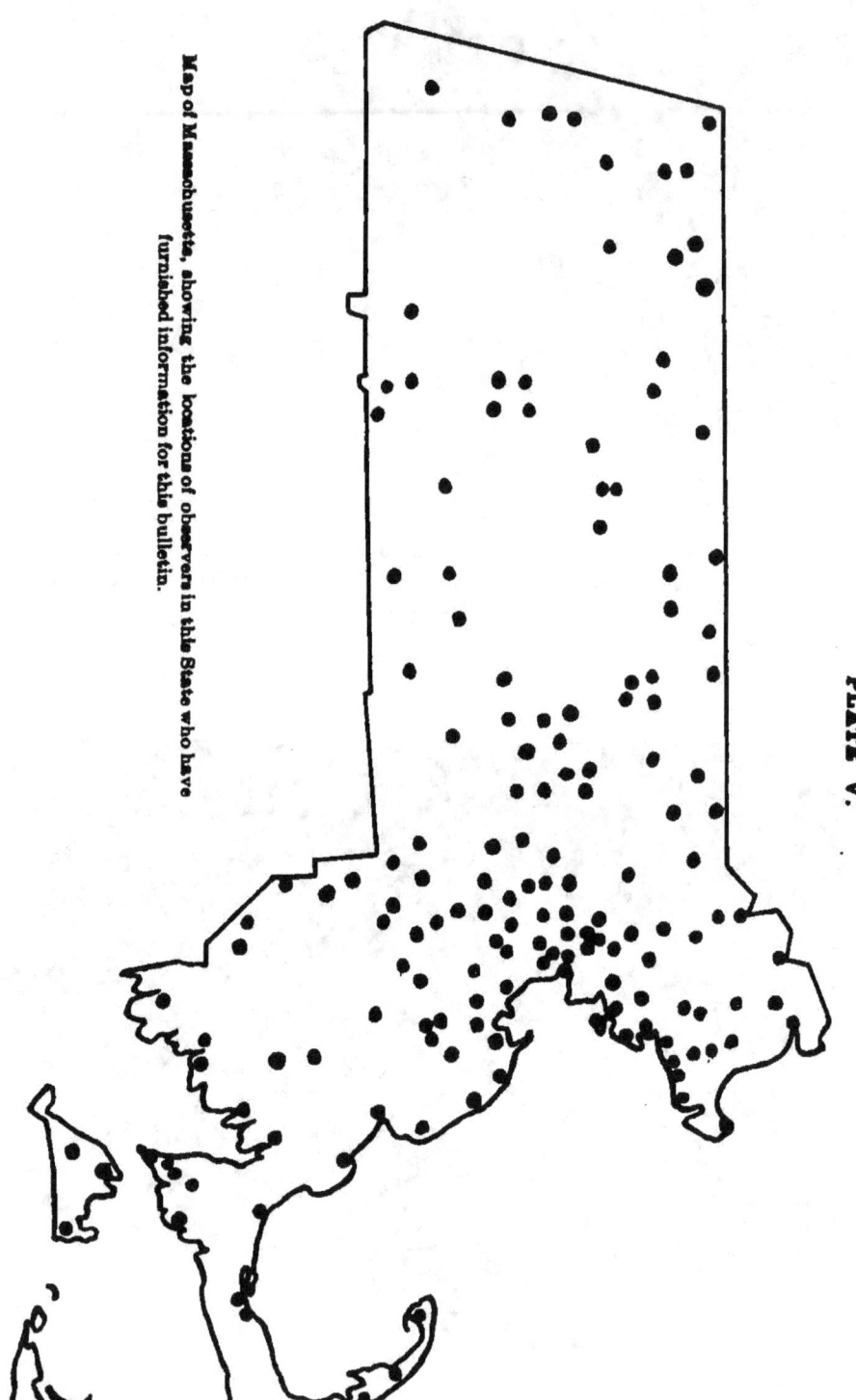

Map of Massachusetts, showing the locations of observers in this State who have furnished information for this bulletin.

PLATE V.

PLATE IV.

SOME ADULT BIRDS BROUGHT IN BY A CAT OR PICKED UP DEAD.

A collection of bird skins in the possession of Miss Cordelia J. Stanwood. Some of these birds were not killed by the cat, but the young birds killed by her were not preserved.

PLATE III.

FIG. 1. — A CAT THAT HAS BEEN "TAUGHT NOT TO KILL BIRDS."

After which she killed them "on the sly." The warbler just killed by her is tied under her chin to "cure" the bird-killing habit, but the expedient failed. She still kills birds.

FIG. 2. — FIFTY-EIGHT BIRDS IN ONE SEASON.

This well-fed pet cat was known to kill fifty-eight birds in one year, including the young in five nests. (Photograph by Mr. A. C. Dike, first published in "Useful Birds.")

PLATE II.

WILD HOUSE CATS KILLED BY MASSACHUSETTS STATE AUTHORITIES TO PROTECT THE HEATH HEN.

Mr. T. Gilbert Pearson examining evidence of numerous wild or stray cats in an uninhabited region. Cats killed on the heath hen reservation on Martha's Vineyard, to preserve this nearly extinct game bird. The region is a wilderness, the nearest villages 3½ and 4 miles away. The only beast of prey on the island is the cat. (Photograph by the courtesy of Dr. G. W. Field.)

PLATE I.

FIG. 1. — VAGABOND HOUSE CAT WITH ROBIN.

The vagabond cat or the barn cat, half-fed or forced to get its own living, becomes a scourge to bird life. Many house cats having once tasted birds or game seem to prefer such food.

FIG. 2. — THE STRAY ALLEY OR ASH BARREL CAT.

Cities and towns radiate such cats, which become very destructive to wild life.

www.ingramcontent.com/pod-product-compliance
Lightning Source LLC
Chambersburg PA
CBHW062217220526
45471CB00009B/3232